Teacher, Student, and Parent
One-Stop Internet Resources

Log on to bookg.msscience.com

ONLINE STUDY TOOLS

- Section Self-Check Quizzes
- Interactive Tutor
- Chapter Review Tests
- Standardized Test Practice
- Vocabulary PuzzleMaker

ONLINE RESEARCH

- WebQuest Projects
- Prescreened Web Links
- Career Links
- Internet Labs

INTERACTIVE ONLINE STUDENT EDITION

- Complete Interactive Student Edition available at mhln.com

FOR TEACHERS

- Teacher Bulletin Board
- Teaching Today—Professional Development

SAFETY SYMBOLS

	HAZARD	EXAMPLES	PRECAUTION	REMEDY
DISPOSAL	Special disposal procedures need to be followed.	certain chemicals, living organisms	Do not dispose of these materials in the sink or trash can.	Dispose of wastes as directed by your teacher.
BIOLOGICAL	Organisms or other biological materials that might be harmful to humans	bacteria, fungi, blood, unpreserved tissues, plant materials	Avoid skin contact with these materials. Wear mask or gloves.	Notify your teacher if you suspect contact with material. Wash hands thoroughly.
EXTREME TEMPERATURE	Objects that can burn skin by being too cold or too hot	boiling liquids, hot plates, dry ice, liquid nitrogen	Use proper protection when handling.	Go to your teacher for first aid.
SHARP OBJECT	Use of tools or glassware that can easily puncture or slice skin	razor blades, pins, scalpels, pointed tools, dissecting probes, broken glass	Practice common-sense behavior and follow guidelines for use of the tool.	Go to your teacher for first aid.
FUME	Possible danger to respiratory tract from fumes	ammonia, acetone, nail polish remover, heated sulfur, moth balls	Make sure there is good ventilation. Never smell fumes directly. Wear a mask.	Leave foul area and notify your teacher immediately.
ELECTRICAL	Possible danger from electrical shock or burn	improper grounding, liquid spills, short circuits, exposed wires	Double-check setup with teacher. Check condition of wires and apparatus.	Do not attempt to fix electrical problems. Notify your teacher immediately.
IRRITANT	Substances that can irritate the skin or mucous membranes of the respiratory tract	pollen, moth balls, steel wool, fiberglass, potassium permanganate	Wear dust mask and gloves. Practice extra care when handling these materials.	Go to your teacher for first aid.
CHEMICAL	Chemicals can react with and destroy tissue and other materials	bleaches such as hydrogen peroxide; acids such as sulfuric acid, hydrochloric acid; bases such as ammonia, sodium hydroxide	Wear goggles, gloves, and an apron.	Immediately flush the affected area with water and notify your teacher.
TOXIC	Substance may be poisonous if touched, inhaled, or swallowed.	mercury, many metal compounds, iodine, poinsettia plant parts	Follow your teacher's instructions.	Always wash hands thoroughly after use. Go to your teacher for first aid.
FLAMMABLE	Flammable chemicals may be ignited by open flame, spark, or exposed heat.	alcohol, kerosene, potassium permanganate	Avoid open flames and heat when using flammable chemicals.	Notify your teacher immediately. Use fire safety equipment if applicable.
OPEN FLAME	Open flame in use, may cause fire.	hair, clothing, paper, synthetic materials	Tie back hair and loose clothing. Follow teacher's instruction on lighting and extinguishing flames.	Notify your teacher immediately. Use fire safety equipment if applicable.

 Eye Safety Proper eye protection should be worn at all times by anyone performing or observing science activities.

 Clothing Protection This symbol appears when substances could stain or burn clothing.

 Animal Safety This symbol appears when safety of animals and students must be ensured.

 Handwashing After the lab, wash hands with soap and water before removing goggles.

The Changing Surface of Earth

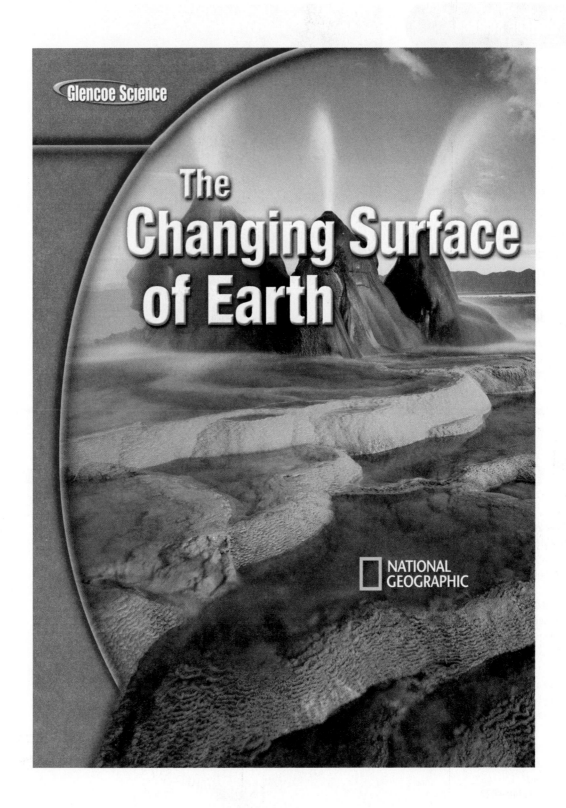

NATIONAL GEOGRAPHIC

Mc Graw Hill **Glencoe**

New York, New York   Columbus, Ohio   Chicago, Illinois   Woodland Hills, California

The Changing Surface of Earth

Fly Geyser, Nevada is a human-made drill well, which is now a constantly spouting hotspring. It is located in the Black Rock Desert, near Gerlach, Nevada. The tufa terraces, or "natural steps," are formed by mineral deposits from the springs.

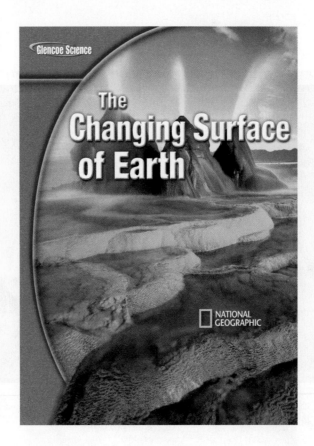

 **Glencoe**

The **McGraw-Hill** Companies

Send all inquiries to:
Glencoe/McGraw-Hill
8787 Orion Place
Columbus, OH 43240-4027

ISBN: 978-0-07-877824-7
MHID: 0-07-877824-7

Printed in the United States of America.

5 6 7 8 9 10 DOW 10

Authors

NATIONAL GEOGRAPHIC
Education Division
Washington, D.C.

Susan Leach Snyder
Retired Teacher/Consultant
Jones Middle School
Upper Arlington, OH

Ralph M. Feather Jr., PhD
Assistant Professor
Department of Educational Studies
and Secondary Education
Bloomsburg University
Bloomsburg, PA

Dinah Zike
Educational Consultant
Dinah-Might Activities, Inc.
San Antonio, TX

Series Consultants

CONTENT

William C. Keel, PhD
Department of Physics and
Astronomy
University of Alabama
Tuscaloosa, AL

Robert Nierste
Science Department Head
Hendrick Middle School, Plano ISD
Plano, TX

MATH

Michael Hopper, DEng
Manager of Aircraft Certification
L-3 Communications
Greenville, TX

Teri Willard, EdD
Mathematics Curriculum Writer
Belgrade, MT

READING

Carol A. Senf, PhD
School of Literature,
Communication, and Culture
Georgia Institute of Technology
Atlanta, GA

SAFETY

Aileen Duc, PhD
Science 8 Teacher
Hendrick Middle School, Plano ISD
Plano, TX

Sandra West, PhD
Department of Biology
Texas State University-San Marcos
San Marcos, TX

ACTIVITY TESTERS

Nerma Coats Henderson
Pickerington Lakeview Jr. High
School
Pickerington, OH

Mary Helen Mariscal-Cholka
William D. Slider Middle School
El Paso, TX

**Science Kit and Boreal
Laboratories**
Tonawanda, NY

Series Reviewers

Lois Burdette
Green Bank Elementary-Middle
School
Green Bank, WV

Marcia Chackan
Pine Crest School
Boca Raton, FL

Karen Curry
East Wake Middle School
Raleigh, NC

Annette D'Urso Garcia
Kearney Middle School
Commerce City, CO

Nerma Coats Henderson
Pickerington Lakeview Jr. High
School
Pickerington, OH

Michael Mansour
Board Member
National Middle Level Science
Teacher's Association
John Page Middle School
Madison Heights, MI

Sharon Mitchell
William D. Slider Middle School
El Paso, TX

HOW TO...

Use Your Science Book

Before You Read

- **Chapter Opener** Science is occurring all around you, and the opening photo of each chapter will preview the science you will be learning about. The **Chapter Preview** will give you an idea of what you will be learning about, and you can try the **Launch Lab** to help get your brain headed in the right direction. The **Foldables** exercise is a fun way to keep you organized.

- **Section Opener** Chapters are divided into two to four sections. The **As You Read** in the margin of the first page of each section will let you know what is most important in the section. It is divided into four parts. **What You'll Learn** will tell you the major topics you will be covering. **Why It's Important** will remind you why you are studying this in the first place! The **Review Vocabulary** word is a word you already know, either from your science studies or your prior knowledge. The **New Vocabulary** words are words that you need to learn to understand this section. These words will be in **boldfaced** print and highlighted in the section. Make a note to yourself to recognize these words as you are reading the section.

Glencoe Science

The
Changing Surface
of Earth

NATIONAL
GEOGRAPHIC

As You Read

- **Headings** Each section has a title in large red letters, and is further divided into blue titles and small red titles at the beginnings of some paragraphs. To help you study, make an outline of the headings and subheadings.

- **Margins** In the margins of your text, you will find many helpful resources. The **Science Online** exercises and **Integrate** activities help you explore the topics you are studying. **MiniLabs** reinforce the science concepts you have learned.

- **Building Skills** You also will find an **Applying Math** or **Applying Science** activity in each chapter. This gives you extra practice using your new knowledge, and helps prepare you for standardized tests.

- **Student Resources** At the end of the book you will find **Student Resources** to help you throughout your studies. These include **Science, Technology,** and **Math Skill Handbooks,** an **English/Spanish Glossary,** and an **Index.** Also, use your **Foldables** as a resource. It will help you organize information, and review before a test.

- **In Class** Remember, you can always ask your teacher to explain anything you don't understand.

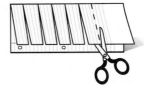

FOLDABLES™
Study Organizer

Science Vocabulary Make the following Foldable to help you understand the vocabulary terms in this chapter.

STEP 1 Fold a vertical sheet of notebook paper from side to side.

STEP 2 Cut along every third line of only the top layer to form tabs.

STEP 3 Label each tab with a vocabulary word from the chapter.

Build Vocabulary As you read the chapter, list the vocabulary words on the tabs. As you learn the definitions, write them under the tab for each vocabulary word.

Look For...

FOLDABLES™

At the beginning of every section.

In Lab

Working in the laboratory is one of the best ways to understand the concepts you are studying. Your book will be your guide through your laboratory experiences, and help you begin to think like a scientist. In it, you not only will find the steps necessary to follow the investigations, but you also will find helpful tips to make the most of your time.

- Each lab provides you with a **Real-World Question** to remind you that science is something you use every day, not just in class. This may lead to many more questions about how things happen in your world.

- Remember, experiments do not always produce the result you expect. Scientists have made many discoveries based on investigations with unexpected results. You can try the experiment again to make sure your results were accurate, or perhaps form a new hypothesis to test.

- Keeping a **Science Journal** is how scientists keep accurate records of observations and data. In your journal, you also can write any questions that may arise during your investigation. This is a great method of reminding yourself to find the answers later.

Look For...
- **Launch Labs** start every chapter.
- **MiniLabs** in the margin of each chapter.
- **Two Full-Period Labs** in every chapter.
- **EXTRA Try at Home Labs** at the end of your book.
- the **Web site** with laboratory demonstrations.

Before a Test

Admit it! You don't like to take tests! However, there *are* ways to review that make them less painful. Your book will help you be more successful taking tests if you use the resources provided to you.

- Review all of the **New Vocabulary** words and be sure you understand their definitions.

- Review the notes you've taken on your **Foldables,** in class, and in lab. Write down any question that you still need answered.

- Review the **Summaries** and **Self Check questions** at the end of each section.

- Study the concepts presented in the chapter by reading the **Study Guide** and answering the questions in the **Chapter Review.**

Look For...

- **Reading Checks** and **caption questions** throughout the text.
- the **Summaries** and **Self Check questions** at the end of each section.
- the **Study Guide** and **Review** at the end of each chapter.
- the **Standardized Test Practice** after each chapter.

Let's Get Started

To help you find the information you need quickly, use the Scavenger Hunt below to learn where things are located in Chapter 1.

1. What is the title of this chapter?

2. What will you learn in Section 1?

3. Sometimes you may ask, "Why am I learning this?" State a reason why the concepts from Section 2 are important.

4. What is the main topic presented in Section 2?

5. How many reading checks are in Section 1?

6. What is the Web address where you can find extra information?

7. What is the main heading above the sixth paragraph in Section 2?

8. There is an integration with another subject mentioned in one of the margins of the chapter. What subject is it?

9. List the new vocabulary words presented in Section 2.

10. List the safety symbols presented in the first Lab.

11. Where would you find a Self Check to be sure you understand the section?

12. Suppose you're doing the Self Check and you have a question about concept mapping. Where could you find help?

13. On what pages are the Chapter Study Guide and Chapter Review?

14. Look in the Table of Contents to find out on which page Section 2 of the chapter begins.

15. You complete the Chapter Review to study for your chapter test. Where could you find another quiz for more practice?

Teacher Advisory Board

Student Advisory Board

The Glencoe middle school science Student Advisory Board taking a timeout at COSI, a science museum in Columbus, Ohio.

Contents

In each chapter, look for these opportunities for review and assessment:
- Reading Checks
- Caption Questions
- Section Review
- Chapter Study Guide
- Chapter Review
- Standardized Test Practice
- Online practice at bookg.msscience.com

Get Ready to Read Strategies
- Questioning 8A
- Make Predictions 36A
- Identify Cause and Effect 64A
- Make Connections 92A
- Take Notes 124A
- Questions and Answers 154A

Contents

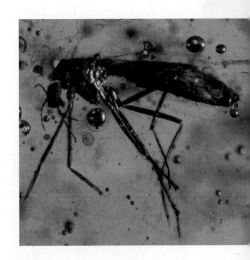

Student Resources

FALLING ROCK

Cross-Curricular Readings/Labs

DVD available as a video lab

Content Details

Labs/Activities

Science Online

Standardized Test Practice

Content Details

Land Use in Floodplains

Imagine what it would be like to watch water rise higher and higher in your house, threatening all your posses-sions. That's what happened to some people in the mid-west in June and July of 1993. The upper Mississippi River basin, with the land already soaked from a wet winter and spring, received almost 14 inches of rain. Even though people struggled desperately to protect their homes and busi-nesses, whole towns were flooded and entire farms were lost, resulting in billions of dollars of damage and loss of life. And yet, after the water receded, the people moved back, homes and businesses were repaired or rebuilt, and new crops were planted.

Living on a Floodplain

Floods are the most common natural disaster in the world and occur when there is persistent heavy rainfall and the soil is water-logged. The excess water swells a river, which eventually spills out onto the surrounding flatlands. Science is limited to offering options instead of solutions to those who choose to live on floodplains—areas prone to floods. It is the task of individuals and government to develop solutions to the problem of flooding.

Figure 1 Major floods devas-tated entire communities, such as Jefferson City, Missouri, during the Great Flood of 1993.

Figure 2 Although rivers usually stay within their chan-nels, rivers cover floodplains when heavy rainfall increases their flow.

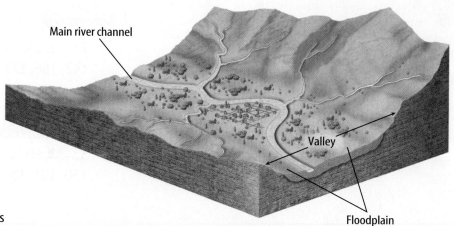

Main river channel

Valley

Floodplain

Almost 10 million homes in the United States are situated on floodplains. When floodplains are not underwater, their flat, fertile soil and proximity to water make them popular sites for towns, farms, and industrial transportation. However, when rivers flow onto floodplains, they also wash through the homes and farms that are located there.

How have scientists and engineers tried to deal with the problems of flooding on developed floodplains? One method is to use human-made structures to block or divert the flow of rivers. Many dams have been built to block water from flowing downstream during heavy rains, forcing it instead into human-made lakes. Thousands of miles of levees, or artificial embankments, have been built along rivers to keep water contained in channels.

Limitations

These efforts have proven to be only partly successful. Human-made systems only can contain and direct water to a certain extent. Dams have overflowed and artificial conduits, or pathways for water, have caused flooding downstream. Levees have caused drying of sponge-like, absorbent wetlands, creating land even more prone to flooding.

Because of these problems, many environmental scientists are in favor of leaving floodplains undeveloped and moving floodplain communities to higher ground with the help of the Federal Emergency Management Agency (FEMA) which provides financial assistance to owners of homes and businesses that have been flooded and wish to relocate. The United States Department of Agriculture's Wetlands Reserve Program also is working to keep floodplain wetlands from being developed and to return developed floodplains to wetlands. The program has successfully restored wetlands in Missouri, Michigan, and several other states.

Figure 3 A wetland can absorb floodwaters more effectively than developed land can.

Figure 4 The floodgates on this dam had to be opened after weeks of heavy rain.

The Limits of Science

Society often asks scientists to solve problems. But there are times when the role of a scientist is limited to offering options. Science deals with facts, but there are questions it cannot answer. For example, people want answers to questions about how to prevent flooding or how to protect the homes, businesses, and farms in floodplains. When answering these questions, however, personal, political and economic factors all must be considered.

Figure 5 A system of levees is designed to keep the Mississippi River from pouring into developed areas.

What Does Science Do?

Science helps people solve problems and answer questions. By experimenting, analyzing data, and forming conclusions, people use science to gain an increased understanding of nature. The procedures used by scientists must be carefully planned to produce reliable results. Scientific methods must be observable, testable, and repeatable.

Scientists can apply scientific methods to the problem of living on floodplains. Identifying the problem of flooding is straightforward. Floodwaters are a danger to people and structures on floodplains. Scientists propose and test various solutions to the problem. Controlling the flow of river water through dams and levees can be beneficial, but it will not completely avoid flooding.

Testing Ideas

Scientists often can perform experiments in a laboratory to test ideas about a problem. For example, they might test how water flows through different types of soil or test the reliability of various levee designs. Other times scientists must rely on observations of real-life situations, and observations of past flooding, in order to reach conclusions. When studying floodplain development, scientists cannot accurately predict how water will flow on developed floodplains.

Sometimes mistakes are made in the process of applying scientific methods to solve a problem. When original hypotheses are proven inaccurate, the experience gained through experimentation can help refine or restructure ideas.

For example, engineers who constructed early dam and levee systems did not do so carelessly. While designing systems to control the flow of water, workers cannot predict all the shortcomings and adverse effects that dams and levees might produce. What is now known about rivers and their floodplains largely comes from earlier attempts at flood control.

What Doesn't Science Do?

Science deals with facts, but it can't tell people how to think or feel. Science exposes the dangers of flooding, but the decision of whether to accept the risks of living on a floodplain remains with those who wish to stay or move there. One solution to the problem of flooding is to leave floodplains undeveloped and to move communities out of floodplains that already have been developed. Science is not qualified to make this decision. Solutions to the problem of flooding must be balanced among different viewpoints.

Moving communities to higher ground removes them from a river—a fertile agricultural site, a scenic area, and perhaps even an industrial work site. By moving entire communities, neighborhoods are lost and historic sites could be destroyed. In addition, people might not wish to lose government funding that maintains development of floodplains.

People ultimately must make the difficult decisions concerning floodplain development. Although science can be applied to help offer possible solutions, it cannot provide the answers to philosophical and political questions that arise from the problem of flooding. Science can make recommendations but it cannot and should not dictate behavior.

Figure 6 This home in Rhineland, Missouri, was moved to higher ground after the Great Flood of 1993.

Think of an important land use decision that your community faces now, or faced recently. Research both sides of the issue. What part did science play in the debate? What about politics and economics? Take a position, and then participate in a class panel discussion of the issue.

Figure 7 These students are using debate as a way to learn about both sides of an issue.

Views of Earth

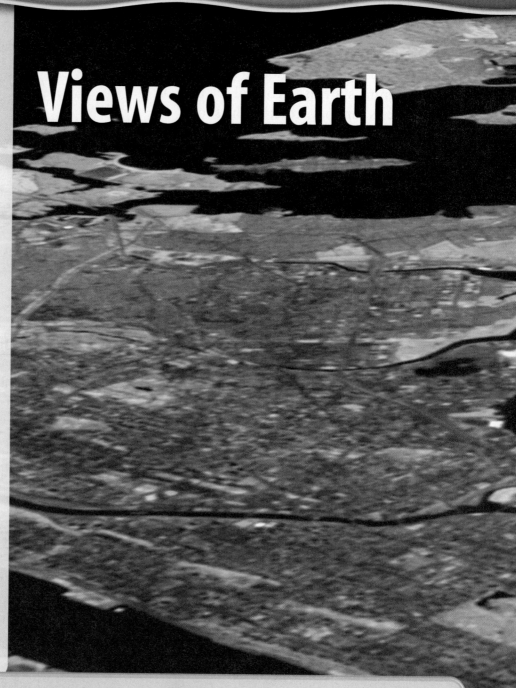

Pictures From Above

Remote sensing from satellites is a powerful way to learn about Earth's landforms, weather, and vegetation. In this image, vegetation shows up as green, uncovered land is red, water is blue, and human-made structures appear gray.

Science Journal Assume that you want to build a home at a location shown somewhere in this photograph. Describe where you would build your new home and why you would build at your chosen location.

Start-Up Activities

Describe Landforms

Pictures of Earth from space are acquired by instruments attached to satellites. Scientists use these images to make maps because they show features of Earth's surface, such as mountains and rivers.

1. Using a globe, atlas, or a world map, locate the following features and describe their positions on Earth relative to other major features.
   a. Andes mountains
   b. Amazon, Ganges, and Mississippi Rivers
   c. Indian Ocean, the Sea of Japan, and the Baltic Sea
   d. Australia, South America, and North America
2. Provide any other details that would help someone else find them.
3. **Think Critically** Choose one country on the globe or map and describe its major physical features in your Science Journal.

Preview this chapter's content and activities at bookg.msscience.com

FOLDABLES
Study Organizer

Views of Earth Make the following Foldable to help identify what you already know, what you want to know, and what you learned about the views of Earth.

STEP 1 Fold a vertical sheet of paper from side to side. Make the front edge about 1.25 cm shorter than the back edge.

STEP 2 Turn lengthwise and **fold** into thirds.

STEP 3 **Unfold and cut** only the top layer along both folds to make three tabs.

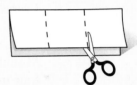

STEP 4 Label each tab.

Know? | Want to know? | Learned?

Identify Questions Before you read the chapter, write what you already know about the views of Earth under the left tab of your Foldable, and write questions about what you want to know under the center tab. After you read the chapter, list what you learned under the right tab.

Get Ready to Read

Questioning

① Learn It! Asking questions helps you to understand what you read. As you read, think about the questions you'd like answered. Often you can find the answer in the next paragraph or section. Learn to ask good questions by asking who, what, when, where, why, and how.

② Practice It! Read the following passage from Section 1.

> Mountains with snowcapped peaks often are shrouded in clouds and tower above the surrounding land. If you climb them, the views are spectacular. The world's highest mountain peak is Mount Everest in the Himalaya—more than 8,800 m above see level. By contrast, the highest mountain peaks in the United States reach just over 6,000 m. Mountains also vary in how they are formed. The four main types of mountains are folded, upwarped, fault-block, and volcanic.
>
> —*from page 11*

Here are some questions you might ask about this paragraph:

• What are the four main types of mountains?
• What causes these four types of mountains to be different?
• Where is the world's highest mountain peak?

③ Apply It! As you read the chapter, look for answers to section headings that are in the form of questions.

Reading Tip

Test yourself. Create questions and then read to find answers to your own questions.

Target Your Reading

Use this to focus on the main ideas as you read the chapter.

① **Before you read** the chapter, respond to the statements below on your worksheet or on a numbered sheet of paper.
- Write an **A** if you **agree** with the statement.
- Write a **D** if you **disagree** with the statement.

② **After you read** the chapter, look back to this page to see if you've changed your mind about any of the statements.
- If any of your answers changed, explain why.
- Change any false statements into true statements.
- Use your revised statements as a study guide.

Science Online

Print out a worksheet of this page at bookg.msscience.com

Before You Read A or D		Statement	After You Read A or D
	1	Plateaus are flat, raised landforms made of nearly horizontal rocks with a steep-sloped boundary.	
	2	Folded mountains are formed by tremendous forces inside Earth squeezing horizontal rock layers.	
	3	Volcanic mountains are cone-shaped structures that formed when molten rock rose to the surface.	
	4	Latitude lines run north to south.	
	5	Latitude lines are also called meridians.	
	6	A map scale is used to measure the weight of heavy maps.	
	7	A map legend is a historic map.	
	8	Contour lines run up and down on hillsides.	
	9	Contour intervals indicate horizontal distance on topographic maps.	
	10	Geologic cross sections can be used to visualize the slope of rock layers beneath Earth's surface.	

Landforms

What You'll Learn

- **Discuss** differences between plains and plateaus.
- **Describe** folded, upwarped, fault-block, and volcanic mountains.

Why It's Important

Landforms influence how people can use land.

⊙ Review Vocabulary

landform: a natural feature of a land surface

New Vocabulary

- plain
- plateau
- folded mountain
- upwarped mountain
- fault-block mountain
- volcanic mountain

Plains

Earth offers abundant variety—from tropics to tundras, deserts to rain forests, and freshwater mountain streams to saltwater tidal marshes. Some of Earth's most stunning features are its landforms, which can provide beautiful vistas, such as vast, flat, fertile plains; deep gorges that cut through steep walls of rock; and towering, snowcapped peaks. **Figure 1** shows the three basic types of landforms—plains, plateaus, and mountains.

Even if you haven't ever visited mountains, you might have seen hundreds of pictures of them in your lifetime. Plains are more common than mountains, but they are more difficult to visualize. **Plains** are large, flat areas, often found in the interior regions of continents. The flat land of plains is ideal for agriculture. Plains often have thick, fertile soils and abundant, grassy meadows suitable for grazing animals. Plains also are home to a variety of wildlife, including foxes, ground squirrels, and snakes. When plains are found near the ocean, they're called coastal plains. Together, interior plains and coastal plains make up half of all the land in the United States.

Figure 1 Three basic types of landforms are plains, plateaus, and mountains.

Plateau

Mountains

Plain

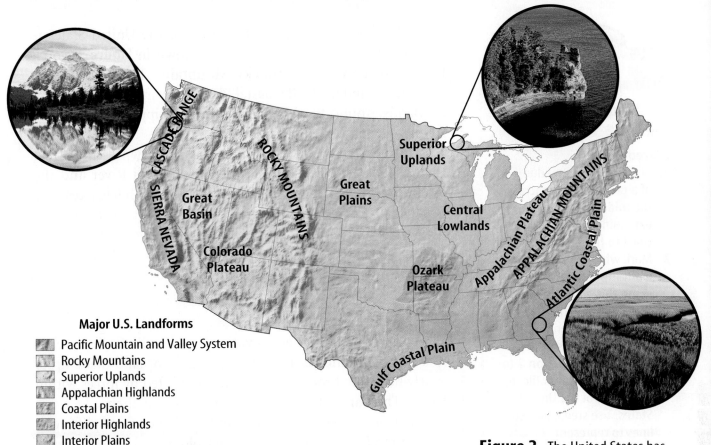

Major U.S. Landforms

- Pacific Mountain and Valley System
- Rocky Mountains
- Superior Uplands
- Appalachian Highlands
- Coastal Plains
- Interior Highlands
- Interior Plains
- Intermontane Plateaus and Basin

Figure 2 The United States has eight major landform regions, which include plains, mountains, and plateaus.
Describe *the region that you live in.*

Coastal Plains A coastal plain often is called a lowland because it is lower in elevation, or distance above sea level, than the land around it. You can think of the coastal plains as being the exposed portion of a continental shelf. The continental shelf is the part of a continent that extends into the ocean. The Atlantic Coastal Plain is a good example of this type of landform. It stretches along the east coast of the United States from New Jersey to Florida. This area has low rolling hills, swamps, and marshes. A marsh is a grassy wetland that usually is flooded with water.

The Atlantic Coastal Plain, shown in **Figure 2,** began forming about 70 million years ago as sediment began accumulating on the ocean floor. Sea level eventually dropped, and the seafloor was exposed. As a result, the coastal plain was born. The size of the coastal plain varies over time. That's because sea level rises and falls. During the last ice age, the coastal plain was larger than it is now because so much of Earth's water was contained in glaciers.

The Gulf Coastal Plain includes the lowlands in the southern United States that surround the Gulf of Mexico. Much of this plain was formed from sediment deposited in deltas by the many rivers that enter the Gulf of Mexico.

Reading Check *How are coastal plains formed?*

Profiling the United States

Procedure

1. Place the bottom edge of a piece of **paper** across the middle of **Figure 2,** extending from the west coast to the east coast.
2. Mark where different landforms are located along this edge.
3. Use a **map of the United States** and the **descriptions of the landforms in Section 1** to help you draw a profile, or side view, of the United States. Use steep, jagged lines to represent mountains. Low, flat lines can represent plains.

Analysis

1. Describe how your profile changed shape as you moved from west to east.
2. Describe how the shape of your profile would be different if you oriented your paper north to south.

Try at Home

Interior Plains The central portion of the United States is comprised largely of interior plains. Shown in **Figure 3,** you'll find them between the Rocky Mountains, the Appalachian Mountains, and the Gulf Coastal Plain. They include the Central Lowlands around the Missouri and Mississippi Rivers and the rolling hills of the Great Lakes area.

A large part of the interior plains is known as the Great Plains. This area lies between the Mississippi River and the Rocky Mountains. It is a flat, grassy, dry area with few trees. The Great Plains also are referred to as the high plains because of their elevation, which ranges from 350 m above sea level at the eastern border to 1,500 m in the west. The Great Plains consist of nearly horizontal layers of sedimentary rocks.

Plateaus

At somewhat higher elevations, you will find plateaus (pla TOHZ). **Plateaus** are flat, raised areas of land made up of nearly horizontal rocks that have been uplifted by forces within Earth. They are different from plains in that their edges rise steeply from the land around them. Because of this uplifting, it is common for plateaus, such as the Colorado Plateau, to be cut through by deep river valleys and canyons. The Colorado River, as shown in **Figure 3,** has cut deeply into the rock layers of the plateau, forming the Grand Canyon. Because the Colorado Plateau is located mostly in what is now a dry region, only a few rivers have developed on its surface. If you hiked around on this plateau, you would encounter a high, rugged environment.

Figure 3 Plains and plateaus are fairly flat, but plateaus have higher elevation.

Mountains

Mountains with snowcapped peaks often are shrouded in clouds and tower high above the surrounding land. If you climb them, the views are spectacular. The world's highest mountain peak is Mount Everest in the Himalaya—more than 8,800 m above sea level. By contrast, the highest mountain peaks in the United States reach just over 6,000 m. Mountains also vary in how they are formed. The four main types of mountains are folded, upwarped, fault-block, and volcanic.

Reading Check *What is the highest mountain peak on Earth?*

Science Online

Topic: Landforms
Visit bookg.msscience.com for Web links to information about some ways landforms affect economic development.

Activity Create four colorful postcards with captions explaining how landforms have affected economic development in your area.

Folded Mountains The Appalachian Mountains and the Rocky Mountains in Canada, shown in **Figure 4,** are comprised of folded rock layers. In **folded mountains,** the rock layers are folded like a rug that has been pushed up against a wall.

INTEGRATE Physics To form folded mountains, tremendous forces inside Earth squeeze horizontal rock layers, causing them to fold. The Appalachian Mountains formed between 480 million and 250 million years ago and are among the oldest and longest mountain ranges in North America. The Appalachians once were higher than the Rocky Mountains, but weathering and erosion have worn them down. They now are less than 2,000 m above sea level. The Ouachita (WAH shuh tah) Mountains of Arkansas are extensions of the same mountain range.

Figure 4 Folded mountains form when rock layers are squeezed from opposite sides. These mountains in Banff National Park, Canada, consist of folded rock layers.

Figure 5 The southern Rocky Mountains are upwarped mountains that formed when crust was pushed up by forces inside Earth.

Upwarped Mountains The Adirondack Mountains in New York, the southern Rocky Mountains in Colorado and New Mexico, and the Black Hills in South Dakota are upwarped mountains. **Figure 5** shows a mountain range in Colorado. Notice the high peaks and sharp ridges that are common to this type of mountain. **Upwarped mountains** form when blocks of Earth's crust are pushed up by forces inside Earth. Over time, the soil and sedimentary rocks at the top of Earth's crust erode, exposing the hard, crystalline rock underneath. As these rocks erode, they form the peaks and ridges.

Fault-Block Mountains **Fault-block mountains** are made of huge, tilted blocks of rock that are separated from surrounding rock by faults. These faults are large fractures in rock along which mostly vertical movement has occurred. The Grand Tetons of Wyoming, shown in **Figure 6,** and the Sierra Nevada in California, are examples of fault-block mountains. As **Figure 6** shows, when these mountains formed, one block was pushed up, while the adjacent block dropped down. This mountain-building process produces majestic peaks and steep slopes.

Figure 6 Fault-block mountains such as the Grand Tetons are formed when faults occur. Some rock blocks move up, and others move down.
Describe *the difference between fault-block mountains and upwarped mountains.*

Figure 7 Mount Shasta is a volcanic mountain made up of layers of lava flows and ash.

Volcanic Mountains **Volcanic mountains,** like the one shown in **Figure 7,** begin to form when molten material reaches the surface through a weak area of the crust. The deposited materials pile up, layer upon layer, until a cone-shaped structure forms. Two volcanic mountains in the United States are Mount St. Helens in Washington and Mount Shasta in California. The Hawaiian Islands are the peaks of huge volcanoes that sit on the ocean floor. Measured from the base, Mauna Loa in Hawaii would be higher than Mount Everest.

Plains, plateaus, and mountains offer different kinds of landforms to explore. They range from low, coastal plains and high, desert plateaus to mountain ranges thousands of meters high.

section 1 review

Summary

Plains and Plateaus

- Plains are large, flat landforms that are usually found in the interior region of a continent.
- Plateaus are flat, raised landforms made of nearly horizontal, uplifted rocks.

Mountains

- Folded mountains form when horizontal rock layers are squeezed from opposite sides.
- Upwarped mountains form when blocks of Earth's crust are pushed up by forces inside Earth.
- Fault-block mountains form from huge, tilted blocks of rock that are separated by faults.
- Volcanic mountains form when molten rock forms cone-shaped structures at Earth's surface.

Self Check

1. **Describe** the eight major landform regions in the United States that are mentioned in this chapter.
2. **Compare and contrast** volcanic mountains, folded mountains, and upwarped mountains using a three-circle Venn diagram.
3. **Think Critically** If you wanted to know whether a particular mountain was formed by movement along a fault, what would you look for? Support your reasoning.

Applying Skills

4. **Concept Map** Make an events-chain concept map to explain how interior plains and coastal plains form.

Viewpoints

What You'll Learn

- **Define** latitude and longitude.
- **Explain** how latitude and longitude are used to identify locations on Earth.
- **Determine** the time and date in different time zones.

Why It's Important

Latitude and longitude allow you to locate places on Earth.

Review Vocabulary
pole: either end of an axis of a sphere

New Vocabulary
- equator
- latitude
- prime meridian
- longitude

Latitude and Longitude

During hurricane season, meteorologists track storms as they form in the Atlantic Ocean. To identify the exact location of a storm, latitude and longitude lines are used. These lines form an imaginary grid system that allows people to locate any place on Earth accurately.

Latitude Look at **Figure 8.** The **equator** is an imaginary line around Earth exactly halfway between the north and south poles. It separates Earth into two equal halves called the northern hemisphere and the southern hemisphere. Lines running parallel to the equator are called lines of **latitude,** or parallels. Latitude is the distance, measured in degrees, either north or south of the equator. Because they are parallel, lines of latitude do not intersect, or cross, one another.

The equator is at 0° latitude, and the poles are each at 90° latitude. Locations north and south of the equator are referred to by degrees north latitude and degrees south latitude, respectively. Each degree is further divided into segments called minutes and seconds. There are 60 minutes in one degree and 60 seconds in one minute.

Figure 8 Latitude and longitude are measurements that are used to indicate locations on Earth's surface.

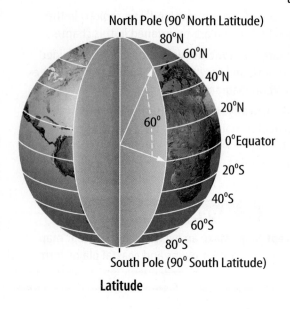

Latitude

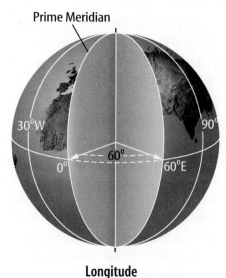

Longitude

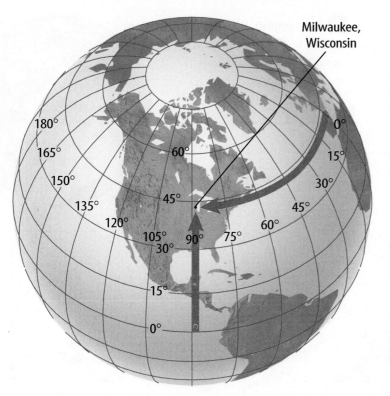

Milwaukee, Wisconsin

Figure 9 The city of Milwaukee, Wisconsin is located at about 43°N, 88°W. **Explain** *the difference between latitude and longitude.*

Longitude The vertical lines, seen in **Figure 8,** have two names—meridians and lines of longitude. Longitude lines are different from latitude lines in many important ways. Just as the equator is used as a reference point for lines of latitude, there's a reference point for lines of longitude—the **prime meridian.** This imaginary line represents 0° longitude. In 1884, astronomers decided the prime meridian should go through the Greenwich (GREN ihtch) Observatory near London, England. The prime meridian had to be agreed upon, because no natural point of reference exists.

Longitude refers to distances in degrees east or west of the prime meridian. Points west of the prime meridian have west longitude measured from 0° to 180°, and points east of the prime meridian have east longitude, measured similarly.

Prime Meridian The prime meridian does not circle Earth as the equator does. Rather, it runs from the north pole through Greenwich, England, to the south pole. The line of longitude on the opposite side of Earth from the prime meridian is the 180° meridian. East lines of longitude meet west lines of longitude at the 180° meridian. You can locate places accurately using latitude and longitude as shown in **Figure 9.** Note that latitude position always comes first when a location is given.

 **Reading Check** *What line of longitude is found opposite the prime meridian?*

Interpreting Latitude and Longitude

Procedure
1. Find the equator and prime meridian on a **world map.**
2. Move your finger to latitudes north of the equator, then south of the equator. Move your finger to longitudes west of the prime meridian, then east of the prime meridian.

Analysis
1. Identify the cities that have the following coordinates:
   a. 56°N, 38°E
   b. 34°S, 18°E
   c. 23°N, 82°W
2. Determine the latitude and longitude of the following cities:
   a. London, England
   b. Melbourne, Australia
   c. Buenos Aires, Argentina

Figure 10 The United States has six time zones.

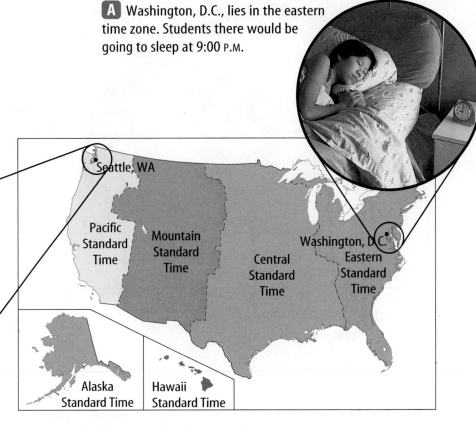

A Washington, D.C., lies in the eastern time zone. Students there would be going to sleep at 9:00 P.M.

B But students in Seattle, Washington, which lies in the Pacific time zone, are eating dinner. **Determine** *what time it would be in Seattle when the students in Washington, D.C., are sleeping at 9:00 P.M.*

Pacific Standard Time

Mountain Standard Time

Central Standard Time

Washington, D.C.
Eastern Standard Time

Seattle, WA

Alaska Standard Time

Hawaii Standard Time

INTEGRATE
Social Studies

International Travel If you travel east or west across three or more time zones, you could suffer from jet lag. Jet lag occurs when your internal time clock does not match the new time zone. Jet lag can disrupt the daily rhythms of sleeping and eating. Have you or any of your classmates ever traveled to a foreign country and suffered from jet lag?

Time Zones

What time it is depends on where you are on Earth. Time is measured by tracking Earth's movement in relation to the Sun. Each day has 24 h, so Earth is divided into 24 time zones. Each time zone is about 15° of longitude wide and is 1 h different from the zones on each side of it. The United States has six different time zones. As you can see in **Figure 10,** people in different parts of the country don't experience dusk simultaneously. Because Earth rotates, the eastern states end a day while the western states are still in sunlight.

Reading Check *What is the basis for dividing Earth into 24 time zones?*

Time zones do not follow lines of longitude strictly. Time zone boundaries are adjusted in local areas. For example, if a city were split by a time zone boundary, the results would be confusing. In such a situation, the time zone boundary is moved outside the city.

Calendar Dates

In each time zone, one day ends and the next day begins at midnight. If it is 11:59 P.M. Tuesday, then 2 min later it will be 12:01 A.M. Wednesday in that particular time zone.

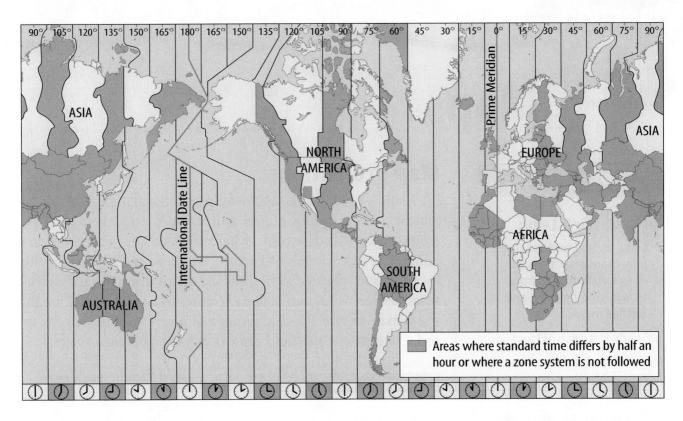

International Date Line You gain or lose time when you enter a new time zone. If you travel far enough, you can gain or lose a whole day. The International Date Line, shown on **Figure 11,** is the transition line for calendar days. If you were traveling west across the International Date Line, located near the 180° meridian, you would move your calendar forward one day. Traveling east, you would move your calendar back one day.

Figure 11 Lines of longitude roughly determine the locations of time zone boundaries. These boundaries are adjusted locally to avoid splitting cities and other political subdivisions, such as counties, into different time zones.

section 2 review

Summary

Latitude and Longitude

- The equator is the imaginary line that wraps around Earth at 0° latitude.
- Latitude is the distance in degrees north or south of the equator.
- The prime meridian is the imaginary line that represents 0° longitude and runs north to south through Greenwich, England.
- Longitude is the distance in degrees east or west of the prime meridian.

Time Zones and Calendar Dates

- Earth is divided into 24 one-hour time zones.
- The International Date Line is the transition line for calendar days.

Self Check

1. **Explain** how lines of latitude and longitude help people find locations on Earth.
2. **Determine** the latitude and longitude of New Orleans, Louisiana.
3. **Calculate** what time it would be in Los Angeles if it were 7:00 P.M. in New York City.
4. **Think Critically** How could you leave home on Monday to go sailing on the ocean, sail for 1 h on Sunday, and return home on Monday?

Applying Math

5. **Use Fractions** If you started at the prime meridian and traveled east one-fourth of the way around Earth, what line of longitude would you reach?

Maps

What **You'll Learn**

- **Compare and contrast** map projections and their uses.
- **Analyze** information from topographic, geologic, and satellite maps.

Why **It's Important**

Maps help people navigate and understand Earth.

Review Vocabulary

globe: a spherical representation of Earth

New Vocabulary

- conic projection
- topographic map
- contour line
- map scale
- map legend

Map Projections

Maps—road maps, world maps, maps that show physical features such as mountains and valleys, and even treasure maps—help you determine where you are and where you are going. They are models of Earth's surface. Scientists use maps to locate various places and to show the distribution of various features or types of material. For example, an Earth scientist might use a map to plot the distribution of a certain type of rock or soil. Other scientists could draw ocean currents on a map.

Reading Check *What are possible uses a scientist would have for maps?*

Many maps are made as projections. A map projection is made when points and lines on a globe's surface are transferred onto paper, as shown in **Figure 12.** Map projections can be made in several different ways, but all types of projections distort the shapes of landmasses or their areas. Antarctica, for instance, might look smaller or larger than it is as a result of the projection that is used for a particular map.

Figure 12 Lines of longitude are drawn parallel to one another in Mercator projections.
Describe *what happens near the poles in Mercator projections.*

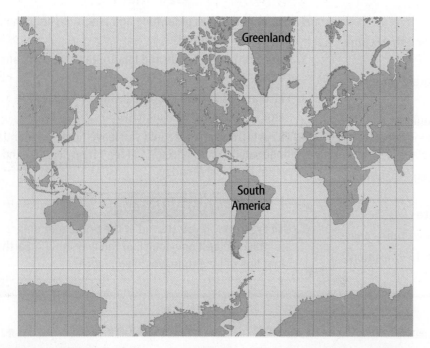

Greenland

South America

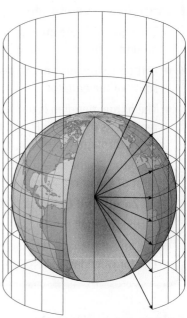

Figure 13 Robinson projections show little distortion in continent shapes and sizes.

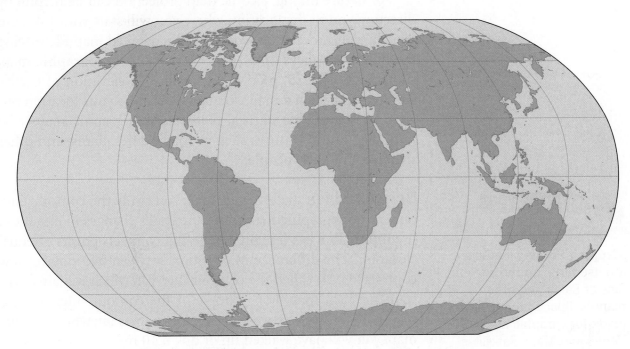

Mercator Projection Mercator (mer KAY ter) projections are used mainly on ships. They project correct shapes of continents, but the areas are distorted. Lines of longitude are projected onto the map parallel to each other. As you learned earlier, only latitude lines are parallel. Longitude lines meet at the poles. When longitude lines are projected as parallel, areas near the poles appear bigger than they are. Greenland, in the Mercator projection in **Figure 12,** appears to be larger than South America, but Greenland is actually smaller.

Robinson Projection A Robinson projection shows accurate continent shapes and more accurate land areas. As shown in **Figure 13,** lines of latitude remain parallel, and lines of longitude are curved as they are on a globe. This results in less distortion near the poles.

Conic Projection When you look at a road map or a weather map, you are using a conic (KAH nihk) projection. Conic projections, like the one shown in **Figure 14,** often are used to produce maps of small areas. These maps are well suited for middle latitude regions but are not as useful for mapping polar or equatorial regions. **Conic projections** are made by projecting points and lines from a globe onto a cone.

Reading Check *How are conic projections made?*

Figure 14 Small areas are mapped accurately using conic projections.

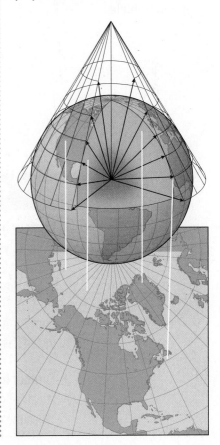

Topographic Maps

For nature hiking, a conic map projection can be helpful by directing you to the location where you will start your hike. On your hike, however, you would need a detailed map identifying the hills and valleys of that specific area. A **topographic map,** shown in **Figure 15,** models the changes in elevation of Earth's surface. With such a map, you can determine your location relative to identifiable natural features. Topographic maps also indicate cultural features such as roads, cities, dams, and other structures built by people.

Contour Lines Before your hike, you study the contour lines on your topographic map to see the trail's changes in elevation. A **contour line** is a line on a map that connects points of equal elevation. The difference in elevation between two side-by-side contour lines is called the contour interval, which remains constant for each map. For example, if the contour interval on a map is 10 m and you walk between two lines anywhere on that map, you will have walked up or down 10 m.

In mountainous areas, the contour lines are close together. This situation models a steep slope. However, if the change in elevation is slight, the contour lines will be far apart. Often large contour intervals are used for mountainous terrain, and small contour intervals are used for fairly flat areas. Why? **Table 1** gives additional tips for examining contour lines.

Index Contours Some contour lines, called index contours, are marked with their elevation. If the contour interval is 5 m, you can determine the elevation of other lines around the index contour by adding or subtracting 5 m from the elevation shown on the index contour.

Mapping Planets Satellites are used to map the surface of Earth and other planets. Space probes have made topographic maps of Venus and Mars. Satellites and probes send a radar beam or laser pulses to the surface and measure how long it takes for the beam or pulses to return to the space vehicle.

Table 1 Contour Rules
1. **Contour lines close around hills and basins.** To decide whether you're looking at a hill or basin, you can read the elevation numbers or look for hachures (ha SHOORZ). These are short lines drawn at right angles to the contour line. They show depressions by pointing toward lower elevations.
2. **Contour lines never cross.** If they did, it would mean that the spot where they cross would have two different elevations.
3. **Contour lines form Vs that point upstream when they cross streams.** This is because streams flow in depressions that are beneath the elevation of the surrounding land surface. When the contour lines cross the depression, they appear as Vs pointing upstream on the map.

NATIONAL GEOGRAPHIC VISUALIZING TOPOGRAPHIC MAPS

Figure 15

Planning a hike? A topographic map will show you changes in elevation. With such a map, you can see at a glance how steep a mountain trail is, as well as its location relative to rivers, lakes, roads, and cities nearby. The steps in creating a topographic map are shown here.

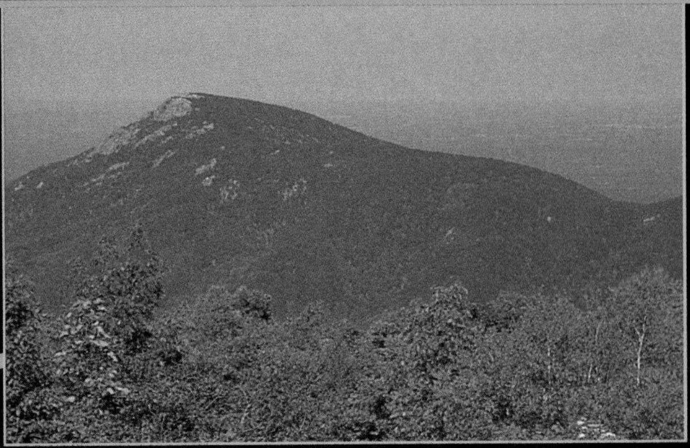

A To create a topographic map of Old Rag Mountain in Shenandoah National Park, Virginia, mapmakers first measure the elevation of the mountain at various points.

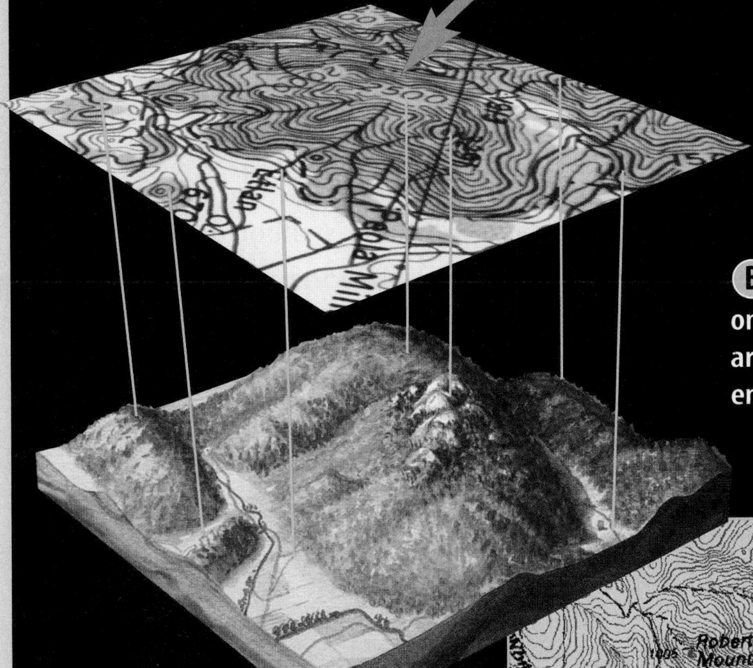

B These points are then projected onto paper. Points at the same elevation are connected, forming contour lines that encircle the mountain.

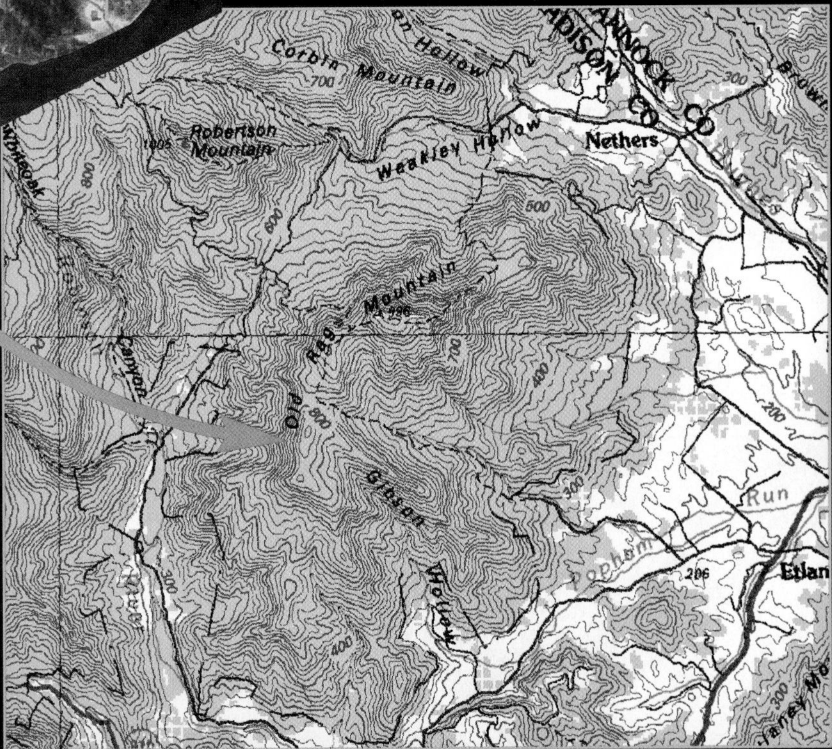

C Where contour lines on a topographic map are close together, elevation is changing rapidly—and the trail is very steep!

Map Scale When planning your hike, you'll want to determine the distance to your destination before you leave. Because maps are small models of Earth's surface, distances and sizes of things shown on a map are proportional to the real thing on Earth. Therefore, real distances can be found by using a scale.

The **map scale** is the relationship between the distances on the map and distances on Earth's surface. Scale often is represented as a ratio. For example, a topographic map of the Grand Canyon might have a scale that reads 1:80,000. This means that one unit on the map represents 80,000 units on land. If the unit you wanted to use was a centimeter, then 1 cm on the map would equal 80,000 cm on land. The unit of distance could be feet or millimeters or any other measure of distance. However, the units of measure on each side of the ratio must always be the same. A map scale also can be shown in the form of a small bar that is divided into sections and scaled down to match real distances on Earth.

Map Legend Topographic maps and most other maps have a legend. A **map legend** explains what the symbols used on the map mean. Some frequently used symbols for topographic maps are shown in the appendix at the back of the book.

Map Series Topographic maps are made to cover different amounts of Earth's surface. A map series includes maps that have the same dimensions of latitude and longitude. For example, one map series includes maps that are 7.5 minutes of latitude by 7.5 minutes of longitude. Other map series include maps covering larger areas of Earth's surface.

Geologic Maps

One of the more important tools to Earth scientists is the geologic map. Geologic maps show the arrangement and types of rocks at Earth's surface. Using geologic maps and data collected from rock exposures, a geologist can infer how rock layers might look below Earth's surface. The block diagram in **Figure 16** is a 3-D model that illustrates a solid section of Earth. The top surface of the block is the geologic map. Side views of the block are called cross sections, which are derived from the surface map. Developing geologic maps and cross sections is extremely important for the exploration and extraction of natural resources. What can a scientist do to determine whether a cross section accurately represents the underground features?

Figure 16 Geologists use block diagrams to understand Earth's subsurface. The different colors represent different rock layers.

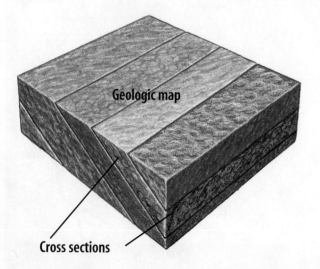

Geologic map

Cross sections

Three-Dimensional Maps Topographic maps and geologic maps are two-dimensional models that are used to study features of Earth's surface. To visualize Earth three dimensionally, scientists often rely on computers. Using computers, information is digitized to create a three-dimensional view of features such as rock layers or river systems. Digitizing is a process by which points are plotted on a coordinate grid.

Map Uses As you have learned, Earth can be viewed in many different ways. Maps are chosen depending upon the situation. If you wanted to determine New Zealand's location relative to Canada and you didn't have a globe, you probably would examine a Mercator projection. In your search, you would use lines of latitude and longitude, and a map scale. If you wanted to travel across the country, you would rely on a road map, or conic projection. You also would use a map legend to help locate features along the way. To climb the highest peak in your region, you would take along a topographic map.

Applying Science

How can you create a cross section from a geologic map?

Earth scientists are interested in knowing the types of rocks and their configurations underground. To help them visualize this, they use geologic maps. Geologic maps offer a two-dimensional view of the three-dimensional situation found under Earth's surface. You don't have to be a professional geologist to understand a geologic map. Use your ability to create graphs to interpret this geologic map.

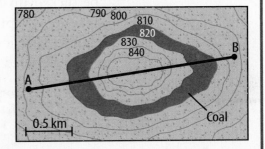

Identifying the Problem

Above is a simple geologic map showing where a coal seam is found on Earth's surface. Place a straight edge of paper along the line marked A–B and mark the points where it meets a contour. Make a different color mark where it meets the exposure of coal. Make a graph on which the various elevations (in meters) are marked on the y-axis. Lay your marked edge of paper along the x-axis and transfer the points directly above onto the proper elevation line. Now connect the dots to draw in the land's surface and connect the marks you made for the coal seam separately.

Solving the Problem

1. What type of topography does the map represent?
2. At what elevation is the coal seam?
3. Does this seam tilt, or is it horizontal? Explain how you know.

Figure 17 Hurricane Isabel's wind lashed the North Carolina and Virginia coasts on September 18, 2003.

Analyze *this satellite photo of Hurricane Isabel approaching the North Carolina Outer Banks. How many states do you think might be affected by this weather system?*

Remote Sensing

Scientists use remote-sensing techniques to collect much of the data used for making maps. Remote sensing is a way of collecting information about Earth from a distance, often using satellites.

Landsat One way that Earth's surface has been studied is with data collected from Landsat satellites, as shown in **Figure 17.** These satellites take pictures of Earth's surface using different wavelengths of light. The images can be used to make maps of snow cover over the United States or to evaluate the impact of forest fires, such as those that occurred in the western United States during the summer of 2000. The newest Landsat satellite, *Landsat 7,* can acquire detailed images by detecting light reflected off landforms on Earth.

Global Positioning System The Global Positioning System, or GPS, is a satellite-based, radio-navigation system that allows users to determine their exact position anywhere on Earth. Twenty-four satellites orbit 20,200 km above the planet. Each satellite sends a position signal and a time signal. The satellites are arranged in their orbits so that signals from at least six can be picked up at any given moment by someone using a GPS receiver. By processing the signals, the receiver calculates the user's exact location. GPS technology is used to navigate, to create detailed maps, and to track wildlife.

section 3 review

Summary

Map Projections
- A map projection is the projection of points and lines of a globe's surface onto paper.

Topographic Maps
- Topographic maps show the changes in elevation of Earth's surface by using contour lines.

Geologic Maps
- Geologic maps show the arrangement and types of rocks at Earth's surface.

Remote Sensing
- Remote sensing is a way of collecting information about Earth from a distance, often by using satellites.
- Distant planets can be mapped using satellites.

Self Check

1. **Compare and contrast** Mercator and conic projections.
2. **Explain** why Greenland appears larger on a Mercator projection than it does on a Robinson projection.
3. **Describe** why contour lines never cross.
4. **Explain** whether a topographic map or a geologic map would be most useful for drilling a water well.
5. **Think Critically** Review the satellite photograph at the beginning of this chapter. Is most of the city near or far from the water? Why is it located there?

Applying Skills

6. **Make Models** Architects make detailed maps called scale drawings to help them plan their work. Make a scale drawing of your classroom.

Making a Tⓐpographic Map

Have you ever wondered how topographic maps are made? Today, radar and remote-sensing devices aboard satellites collect data, and computers and graphic systems make the maps. In the past, surveyors and aerial photographers collected data. Then, maps were hand drawn by cartographers, or mapmakers. In this lab, you can practice cartography.

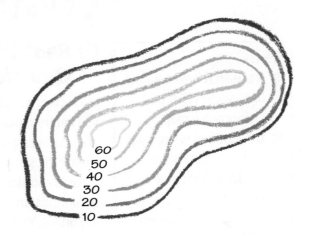

▶ Real-World Question

How is a topographic map made?

Goals
- ■ **Draw** a topographic map.
- ■ **Compare and contrast** contour intervals.

Materials
plastic model of a landform
water tinted with food coloring
transparency
clear-plastic storage box with lid
beaker
metric ruler
tape
transparency marker

▶ Procedure

1. Using the ruler and the transparency marker, make marks up the side of the storage box that are 2 cm apart.

2. Secure the transparency to the outside of the box lid with tape.

3. Place the plastic model in the box. The bottom of the box will be zero elevation.

4. Using the beaker, pour water into the box to a height of 2 cm. Place the lid on the box.

5. Use the transparency marker to trace the top of the water line on the transparency.

6. Using the scale 2 cm = 10 m, mark the elevation on the line.

7. Repeat the process of adding 2 cm of water and tracing until the landform is mapped.

8. Transfer the tracing of the landform onto a sheet of white paper.

▶ Conclude and Apply

1. **Identify** the contour interval of this topographic map.

2. **Evaluate** how the distance between contour lines on the map shows the steepness of the slope on the landform model.

3. **Determine** the total elevation of the landform you have selected.

4. **Describe** how elevation was represented on your map.

5. **Explain** how elevations are shown on topographic maps.

6. Must all topographic maps have a contour line that represents 0 m of elevation? Explain.

Model and Invent

Constructing Landfors

Goals

■ **Research** how contour lines show relief on a topographic map.

■ **Determine** what scale you can best use to model a landscape of your choice.

■ Working cooperatively with your classmates, model a landscape in three dimensions from the information given on a topographic map.

Possible Materials

U.S. Geological Survey 7.5-minute quadrangle maps

sandbox sand

rolls of brown paper towels

spray bottle filled with water

ruler

⊙ Real-World Question

Most maps perform well in helping you get from place to place. A road map, for example, will allow you to choose the shortest route from one place to another. If you are hiking, though, distance might not be so important. You might want to choose a route that avoids steep terrain. In this case you need a map that shows the highs and lows of Earth's surface, called relief. Topographic maps use contour lines to show the landscape in three dimensions. Among their many uses, such maps allow hikers to choose routes that maximize the scenery and minimize the physical exertion. What does a landscape depicted on a two-dimensional topographic map look like in three dimensions? How can you model a landscape?

⊙ Make a Model

1. **Choose** a topographic map showing a landscape easily modeled using sand. Check to see what contour interval is used on the map. Use the index contours to find the difference between the lowest and the highest elevations shown on the landscape. Check the distance scale to determine how much area the landscape covers.

2. **Determine** the scale you will use to convert the elevations shown on your map to heights on your model. Make sure the scale is proportional to the distances on your map.

3. **Plan** a model of the landscape in sand by sketching the main features and their scaled heights onto paper. Note the degree of steepness found on all sides of the features.

4. **Prepare** a document that shows the scale you plan to use for your model and the calculations you used to derive that scale. Remember to use the same scale for distance as you use for height. If your landscape is fairly flat, you can exaggerate the vertical scale by a factor of two or three. Be sure your paper is neat, is easy to follow, and includes all units. Present the document to your teacher for approval.

ⓦ *Test Your Model*

1. Using the sand, spray bottle, and ruler, create a scale model of your landscape on the brown paper towels.

2. **Check** your topographic map to be sure your model includes the landscape features at their proper heights and proper degrees of steepness.

ⓦ *Analyze Your Data*

1. **Determine** if your model accurately represents the landscape depicted on your topographic map. Discuss the strengths and weaknesses of your model.

2. **Explain** why it was important to use the same scale for height and distance. If you exaggerated the height, why was it important to indicate the exaggeration on your model?

ⓦ *Conclude and Apply*

1. **Infer** why the mapmakers chose the contour interval used on your topographic map?

2. **Predict** the contour intervals mapmakers might choose for topographic maps of the world's tallest mountains— the Himalaya—and for topographic maps of Kansas, which is fairly flat.

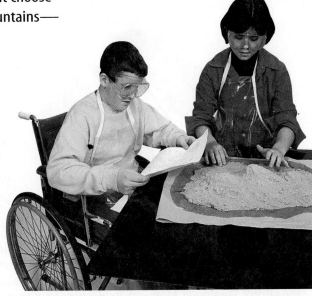

𝒞ommunicating
Your Data

Prepare a vacation getaway commercial to advertise the topographical features of your model landscape. Be sure to discuss the landscape elevation and features, scale, and similarities to actual landforms.

New York Harbor in 1849

Rich Midwest farmland

Georgia peaches

Alaska pipeline

Maine fishing and lobster industry

LOCATION, LOCATION

hy is New York City at the mouth of the Hudson River and not 300 km inland? Why are there more farms in Iowa than in Alaska? What's the reason for growing lots of peaches in Georgia but not in California's Death Valley? It's all about location. The landforms, climate, soil, and resources in an area determine where cities and farms grow and what people connected with them do.

LANDFORMS ARE KEY

When many American cities were founded hundreds of years ago, waterways were the best means of transportation. Old cities such as New York City and Boston are located on deep harbors where ships could land with people and goods. Rivers also were major highways centuries ago. They still are.

Topography and soil also play a role in where activities such as farming take root. States such as Iowa and Illinois have many farms because they have flat land and fertile soil. Growing crops is more difficult in mountainous areas or where soil is stony and poor.

CLIMATE AND SOIL

Climate limits the locations of cities and farms, as well. The fertile soil and warm, moist climate of Georgia make it a perfect place to grow peaches. California's Death Valley can't support such crops because it's a hot, dry desert.

RESOURCES RULE

The location of an important natural resource can change the rules. A gold deposit or an oil field can cause a town to grow in a place where the topography, soil, and climate are not favorable. For example, thousands of people now live in parts of Alaska only because of the great supply of oil there. Maine has a harsh climate and poor soil. But people settled along its coast because they could catch lobsters and fish in the nearby North Atlantic.

The rules that govern where towns grow and where people live are different now than they used to be. Often information, not goods, moves from place to place on computers that can be anywhere. But as long as people farm, use minerals, and transport goods from place to place, the natural environment and natural resources will always help determine where people are and what they do.

Research Why was your community built where it is? Research its history. What types of economic activity were important when it was founded? Did topography, climate, or resources determine its location? Design a Moment in History to share your information.

Science online

For more information, visit
bookg.msscience.com/time

Reviewing Main Ideas

Section 1 Landforms

1. The three main types of landforms are plains, plateaus, and mountains.

2. Plains are large, flat areas. Plateaus are relatively flat, raised areas of land made up of nearly horizontal rocks that have been uplifted. Mountains rise high above the surrounding land.

Section 2 Viewpoints

1. Latitude and longitude form an imaginary grid system that enables points on Earth to be located exactly.

2. Latitude is the distance in degrees north or south of the equator. Longitude is the distance in degrees east or west of the prime meridian.

3. Earth is divided into 24 time zones. Each time zone represents a 1-h difference. The International Date Line separates different calendar days.

Section 3 Maps

1. Mercator, Robinson, and conic projections are made by transferring points and lines on a globe's surface onto paper.

2. Topographic maps show the elevation of Earth's surface. Geologic maps show the types of rocks that make up Earth's surface.

3. Remote sensing is a way of collecting information from a distance. Satellites are important remote-sensing devices.

Visualizing Main Ideas

Copy and complete the following concept map on landforms.

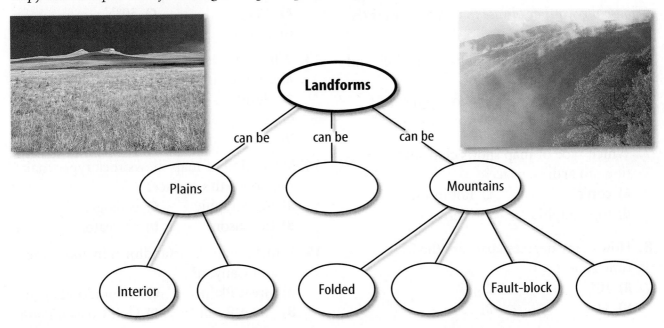

Using Vocabulary

conic projection p. 19	map scale p. 22
contour line p. 20	plain p. 8
equator p. 14	plateau p. 10
fault-block	prime meridian p. 15
mountain p. 12	topographic map p. 20
folded mountain p. 11	upwarped
latitude p. 14	mountain p. 12
longitude p. 15	volcanic mountain p. 13
map legend p. 22	

For each set of terms below, choose the one term that does not belong and explain why it does not belong.

1. upwarped mountain—equator—volcanic mountain

2. plain—plateau—prime meridian

3. topographic map—contour line—volcanic mountain

4. prime meridian—equator—folded mountain

5. fault-block mountain—upwarped mountain—plateau

Checking Concepts

Choose the word or phrase that best answers the question.

6. What makes up about 50 percent of all land areas in the United States?
   - **A)** plateaus
   - **B)** plains
   - **C)** mountains
   - **D)** volcanoes

7. Which type of map shows changes in elevation at Earth's surface?
   - **A)** conic
   - **B)** topographic
   - **C)** Robinson
   - **D)** Mercator

8. How many degrees apart are the 24 time zones?
   - **A)** 10°
   - **B)** 34°
   - **C)** 15°
   - **D)** 25°

Use the photo below to answer question 9.

9. What kind of mountains are the Grand Tetons of Wyoming?
   - **A)** fault-block
   - **B)** volcanic
   - **C)** upwarped
   - **D)** folded

10. Landsat satellites collect data by using
    - **A)** sonar.
    - **B)** echolocation.
    - **C)** sound waves.
    - **D)** light waves.

11. Which type of map is most distorted at the poles?
    - **A)** conic
    - **B)** topographic
    - **C)** Robinson
    - **D)** Mercator

12. Where is the north pole located?
    - **A)** 0°N
    - **B)** 180°N
    - **C)** 50°N
    - **D)** 90°N

13. What is measured with respect to sea level?
    - **A)** contour interval
    - **B)** elevation
    - **C)** conic projection
    - **D)** sonar

14. What kind of map shows rock types making up Earth's surface?
    - **A)** topographic
    - **B)** Robinson
    - **C)** geologic
    - **D)** Mercator

15. Which major U.S. landform includes the Grand Canyon?
    - **A)** Great Plains
    - **B)** Great Basin
    - **C)** Colorado Plateau
    - **D)** Gulf Coastal Plain

Science Online bookg.msscience.com/vocabulary_puzzlemaker

Thinking Critically

16. Explain how a topographic map of the Atlantic Coastal Plain differs from a topographic map of the Rocky Mountains.

17. Determine If you left Korea early Wednesday morning and flew to Hawaii, on what day of the week would you arrive?

Use the illustration below to answer question 18.

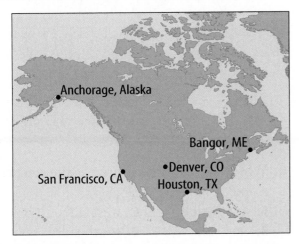

18. Determine Using the map above, arrange these cities in order from the city with the earliest time to the one with the latest time on a given day: Anchorage, Alaska; San Francisco, California; Bangor, Maine; Denver, Colorado; Houston, Texas.

19. Describe how a map with a scale of 1:50,000 is different from a map with a scale of 1:24,000.

20. Compare and contrast Mercator, Robinson, and conic map projections.

21. Form Hypotheses You are visiting a mountain in the northwest part of the United States. The mountain has steep sides and is not part of a mountain range. A crater can be seen at the top of the mountain. Hypothesize about what type of mountain you are visiting.

22. Concept Map Copy and complete the following concept map about parts of a topographic map.

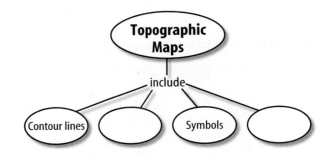

Performance Activities

23. Poem Create a poem about one type of landform. Include characteristics of the landform in your poem. How can the shape of your poem add meaning to your poem? Display your poem with those of your classmates.

Applying Math

24. Calculate If you were flying directly south from the north pole and reached 70° north latitude, how many more degrees of latitude would you pass over before reaching the south pole? Illustrate and label a diagram to support your answer.

Use the map below to answer question 25.

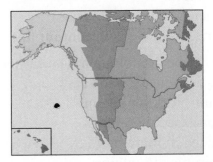

25. Calculate If it is 2:00 P.M. in Orlando, Florida, what time is it in Los Angeles, California? In Anchorage, Alaska? In your hometown?

Record your answers on the answer sheet provided by your teacher or on a sheet of paper.

Use the map below to answer question 1.

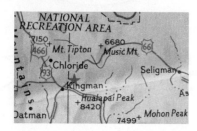

1. Which of the following is shown above?
   A. cross section   **C.** topographic map
   B. geologic map   **D.** road map

2. Which landform is a relatively flat area that has high elevation?
   A. mountain   **C.** coastal plain
   B. interior plain   **D.** plateau

3. Which of the following can provide detailed information about your position on Earth's surface?
   A. prime meridian
   B. global positioning system
   C. International Date Line
   D. LandSat 7

4. What connects points of equal elevation on a map?
   A. legend   **C.** scale
   B. series   **D.** contour line

5. Which type of mountain forms when rock layers are squeezed and bent?
   A. fault-block mountains
   B. upwarped mountains
   C. folded mountains
   D. volcanic mountains

Test-Taking Tip

Read Carefully Read all choices before answering the questions.

Use the illustration below to answer questions 6–8. The numbers on the drawing represent meters above sea level.

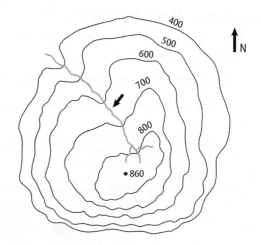

6. Which side of the feature has the steepest slope?
   A. north side   **C.** west side
   B. east side   **D.** south side

7. What is the highest elevation on the feature?
   A. 800 meters   **C.** 860 meters
   B. 960 meters   **D.** 700 meters

8. What does the line that is marked by the arrow represent?
   A. a contour line   **C.** a high ridge
   B. a stream   **D.** a glacier

9. Which type of map is made by projecting points and lines from a globe onto a cone?
   A. Mercator projection
   B. conic projection
   C. Robinson projection
   D. geologic map

10. Which are useful for measuring position north or south of the equator?
    A. lines of latitude
    B. lines of longitude
    C. index contours
    D. map legends

Part 2 | Short Response/Grid In

Record your answers on the answer sheet provided by your teacher or on a separate sheet of paper.

11. How do volcanic mountains form?

12. What is a time zone? How are time zones determined around the world? Why are they needed?

13. Which type of map would you use to find the location of a layer of coal at Earth's surface? Why?

14. How are plateaus similar to plains? How are they different?

15. Why are lines of latitude sometimes called parallels?

16. List the locations on Earth that represent 0° latitude, 90°N latitude, and 90°S latitude.

Use the scale at the bottom of the map to answer questions 17–19.

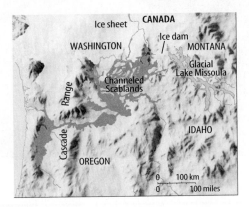

17. How many centimeters on Earth's surface are represented by one centimeter on the map?

18. How many meters on Earth's surface does one centimeter represent?

19. How many kilometers on Earth's surface does one centimeter represent?

Part 3 | Open Ended

Record your answers on a sheet of paper.

Use the map below to answer questions 20–21.

20. Where do mountains, plateaus, and plains occur in the United States? Describe these regions and how they were formed.

21. Take the role of a fur trader, pioneer, or explorer. Write three journal entries that give a general description of how the landforms change across the United States.

22. Compare and contrast interior plains and coastal plains.

23. Why is remote sensing important to society? What types of information are obtained?

24. Why are computers often used to make maps?

25. Why might you experience jet lag if you travel across the United States on a plane? Predict which direction it would be more difficult to adjust to jet lag—when traveling from Hawaii to New York or New York to Hawaii. Support your answer with an example of each situation.

The BIG Idea

Soil is a natural resource that must be monitored, managed, and protected.

SECTION 1
Weathering

Main Idea Weathering processes weaken and break apart rock material into smaller pieces.

SECTION 2
The Nature of Soil

Main Idea Soil is a mixture of weathered rock, decayed organic matter, mineral fragments, water, and air.

SECTION 3
Soil Erosion

Main Idea Soil erosion is harmful because plants do not grow as well when top-soil has been removed.

Weathering and Soil

What's a tor?

A tor, shown in the photo, is a pile of boulders left on the land. Tors form because of weathering, which is a natural process that breaks down rock. Weathering weakened the rock that used to be around the boulders. This weakened rock then was eroded away, and the boulders are all that remain.

Science Journal Write a poem about a tor. Use words in your poem that rhyme with the word *tor*.

Start-Up Activities

Stalactites and Stalagmites

During weathering, minerals can be dissolved by acidic water. If this water seeps into a cave, minerals might precipitate. In this lab, you will model the formation of stalactites and stalagmites.

1. Pour 700 mL of water into two 1,000-mL beakers and place the beakers on a large piece of cardboard. Stir Epsom salt into each beaker until no more will dissolve.

2. Add two drops of yellow food coloring to each beaker and stir.

3. Measure and cut three 75-cm lengths of cotton string. Hold the three pieces of string in one hand and twist the ends of all three pieces to form a loose braid of string.

4. Tie each end of the braid to a large steel nut.

5. Soak the braid of string in one of the beakers until it is wet with the solution. Drop one nut into one beaker and the other nut into the second beaker. Allow the string to sag between the beakers. Observe for several days.

6. **Think Critically** Record your observations in your Science Journal. How does this activity model the formation of stalactites and stalagmites?

Weathering and Soil Make the following Foldable to help you understand the vocabulary terms in this chapter.

STEP 1 Fold a vertical sheet of notebook paper from side to side.

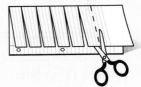

STEP 2 Cut along every third line of only the top layer to form tabs.

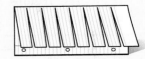

STEP 3 Label each tab.

Build Vocabulary As you read the chapter, list the vocabulary words about weathering and soil on the tabs. As you learn the definitions, write them under the tab for each vocabulary word.

Science Online Preview this chapter's content and activities at bookg.msscience.com

Get Ready to Read

Make Predictions

1 Learn It! A prediction is an educated guess based on what you already know. One way to predict while reading is to guess what you believe the author will tell you next. As you are reading, each new topic should make sense because it is related to the previous paragraph or passage.

2 Practice It! Read the excerpt below from section. Based on what you have read, make predictions about what you will read in the rest of the lesson. After you read section, go back to your predictions to see if they were correct.

> **Think about how you would describe different climates in different regions on Earth.**

> **Predict some different types of soils for different places. What factors might produce different types of soils?**

> **Determine how soil temperature and moisture content could affect the quality of soils.**

Different regions on Earth have **different climates.** Deserts are dry, prairies are semidry, and temperate forests are mild and moist. These places also have **different types of soils. Soil temperature and moisture content** affect the quality of soils. Soils in deserts contain little organic material and are thinner than soils in wetter climates. Prairie soils have thick, dark A horizons because the grasses that grow there contribute lots of organic matter.

—from page 47

3 Apply It! Before you read, skim the questions in the Chapter Review. Choose three questions and predict the answers.

Reading Tip

As you read, check the predictions you made to see if they were correct.

Target Your Reading

Use this to focus on the main ideas as you read the chapter.

1 **Before you read** the chapter, respond to the statements below on your worksheet or on a numbered sheet of paper.

- Write an **A** if you **agree** with the statement.
- Write a **D** if you **disagree** with the statement.

2 **After you read** the chapter, look back to this page to see if you've changed your mind about any of the statements.

- If any of your answers changed, explain why.
- Change any false statements into true statements.
- Use your revised statements as a study guide.

Science Online

Print out a worksheet of this page at bookg.msscience.com

Before You Read A or D		Statement	After You Read A or D
	1	Weathering breaks rocks into smaller and smaller pieces, such as sand, silt, or clay.	
	2	Exposure to atmospheric water and gases causes rocks to change chemically.	
	3	Soil is a mixture of weathered rock, decayed organic matter, mineral fragments, water, and air.	
	4	Because of weathering, new soil is usually produced rapidly in all regions on Earth.	
	5	The different layers of soil are called horizons.	
	6	Climate does not affect the type of soil produced in Earth's different regions.	
	7	Most plants grow well when topsoil erodes.	
	8	In tropical, deforested areas, soil is useful to farmers for only a few years before the topsoil is gone.	
	9	Contour farming is a practice of planting crops in large, circular mounds.	

Weathering

What **You'll Learn**

- **Explain** how mechanical weathering and chemical weathering differ.
- **Describe** how weathering affects Earth's surface.
- **Explain** how climate affects weathering.

Why **It's Important**

Through time, weathering turns mountains into sediment.

🔊 **Review Vocabulary**

surface area: the area of a rock or other object that is exposed to the surroundings

New Vocabulary

- weathering
- mechanical weathering
- ice wedging
- chemical weathering
- oxidation
- climate

Weathering and Its Effects

Can you believe that tiny moss plants, a burrowing vole shrew, and even oxygen in the air can affect solid rock? These things weaken and break apart rock at Earth's surface. Surface processes that work to break down rock are called **weathering.**

Weathering breaks rock into smaller and smaller pieces, such as sand, silt, and clay. These particles are called sediment. The terms *sand, silt,* and *clay* are used to describe specific particle sizes, which contribute to soil texture. Sand grains are larger than silt, and silt is larger than clay.

Soil texture influences virtually all mechanical and chemical processes in the soil, including the ability to hold moisture and nutrients.

Over millions of years, weathering has changed Earth's surface. The process continues today. Weathering wears mountains down to hills, as shown in **Figure 1.** Rocks at the top of mountains are broken down by weathering, and the sediment is moved downhill by gravity, water, and ice. Weathering also produces strange rock formations like those shown at the beginning of this chapter. Two different types of weathering—mechanical weathering and chemical weathering—work together to shape Earth's surface.

Figure 1 Over long periods of time, weathering wears mountains down to rolling hills.
Explain *how this occurs.*

Figure 2 Growing tree roots can be agents of mechanical weathering.

Tree roots can crack a sidewalk.

Tree roots also can grow into cracks and break rock apart.

Mechanical Weathering

Mechanical weathering occurs when rocks are broken apart by physical processes. This means that the overall chemical makeup of the rock stays the same. Each fragment has characteristics similar to the original rock. Growing plants, burrowing animals, and expanding ice are some of the things that can mechanically weather rock. These physical processes produce enough force to break rocks into smaller pieces.

Reading Check *What can cause mechanical weathering?*

Figure 3 Small animals mechanically weather rock when they burrow by breaking apart sediment.

Plants and Animals Water and nutrients that collect in the cracks of rocks result in conditions in which plants can grow. As the roots grow, they enlarge the cracks. You've seen this kind of mechanical weathering if you've ever tripped on a crack in a sidewalk near a tree, as shown in **Figure 2.** Sometimes tree roots wedge rock apart, also shown in **Figure 2.**

Burrowing animals also cause mechanical weathering, as shown in **Figure 3.** As these animals burrow, they loosen sediment and push it to the surface. Once the sediment is brought to the surface, other weathering processes act on it.

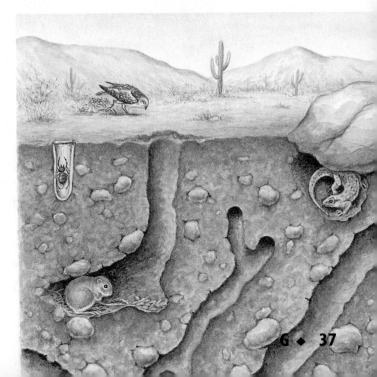

Figure 4 When water enters cracks in rock and freezes, it expands, causing the cracks to enlarge and the rock to break apart.

Figure 5 As rock is broken apart by mechanical weathering, the amount of rock surface exposed to air and water increases. The background squares show the total number of surfaces exposed.

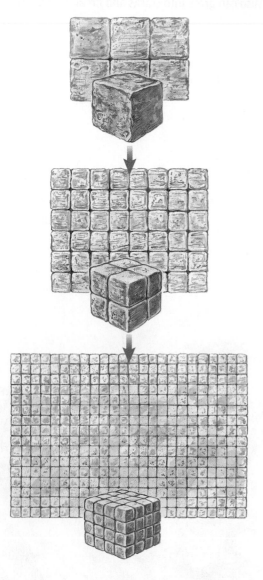

Ice Wedging A mechanical weathering process called ice wedging is shown in **Figure 4. Ice wedging** occurs in temperate and cold climates where water enters cracks in rocks and freezes. Because water expands when it turns to ice, pressure builds up in the cracks. This pressure can extend the cracks and break apart rock. The ice then melts, allowing more water to enter the crack, where it freezes and breaks the rock even more. Ice wedging is most noticeable in the mountains, where warm days and cold nights are common. It is one process that wears down mountain peaks. This cycle of freezing and thawing not only breaks up rocks, but also can break up roads and highways. When water enters cracks in road pavement and freezes, it forces the pavement apart. This causes potholes to form in roads.

Surface Area Mechanical weathering by plants, animals, and ice wedging reduces rocks to smaller pieces. These small pieces have more surface area than the original rock body, as shown in **Figure 5.** As the amount of surface area increases, more rock is exposed to water and oxygen. This speeds up a different type of weathering called chemical weathering, which continues to reduce the particle size of sediments from a coarse to a finer texture.

Chemical Weathering

The second type of weathering, **chemical weathering,** occurs when chemical reactions dissolve or alter the minerals in rocks or change them into different minerals. This type of weathering occurs at or near Earth's surface and changes the chemical composition of the rock, which can weaken the rock.

Natural Acids Naturally formed acids can weather rocks. When water reacts with carbon dioxide in the air or soil, a weak acid, called carbonic acid, forms. Carbonic acid reacts with minerals such as calcite, which is the main mineral that makes up limestone. This reaction causes the calcite to dissolve. Over many thousands of years, carbonic acid has weathered so much limestone that caves have formed, as shown in **Figure 6**.

Chemical weathering also occurs when naturally formed acids come in contact with other rocks. Over a long period of time, the mineral feldspar, which is found in granite, some types of sandstone, and other rocks, is broken down into a clay mineral called kaolinite (KAY oh luh nite). Kaolinite clay is common in some soils. Clay is an end product of weathering.

Reading Check *How does kaolinite clay form?*

Plant Acids Some roots and decaying plants give off acids that also dissolve minerals in rock. When the minerals dissolve, the rock is weakened. Eventually, the rock breaks into smaller pieces. As the rock weathers, nutrients become available to plants.

Science Online

Topic: Chemical Weathering
Visit bookg.msscience.com for Web links to information about chemical weathering.

Activity List different types of chemical weathering. Next to each type, write an effect that you have observed.

Figure 6 Caves form when slightly acidic groundwater dissolves limestone.
Explain *why the groundwater is acidic.*

Carbon dioxide + Water = Carbonic acid
Carbonic acid dissolves limestone.

Observing Chemical and Mechanical Weathering

Procedure

1. Collect and rinse two hand-fuls of **common rock or shells.**
2. Place equal amounts of rock into two **plastic bottles.**
3. Fill one bottle with **water** to cover the rock and seal with a **lid.**
4. Cover the rock in the second bottle with **lemon juice** and seal.
5. Shake both bottles for ten minutes.
6. Tilt the bottles so you can observe the liquids in each.

Analysis
1. Describe the appearance of each liquid.
2. Explain any differences.

Try at Home

Figure 7 Iron-containing minerals like the magnetite shown here can weather to form a rustlike material called limonite.
Explain *how this is similar to rust forming on your bicycle chain.*

pH Scale The strength of acids and bases is measured on the pH scale with a range of 0 to 14. On this scale, 0 is extremely acidic, 14 is extremely basic or alkaline, and 7 is neutral. Most minerals are more soluble in acidic soils than in neutral or slightly alkaline soils. Different plants grow best at different pH values. For example, peanuts grow best in soils that have a pH of 5.3 to 6.6, while alfalfa grows best in soils having a pH of 6.2 to 7.8.

Oxygen Oxygen also causes chemical weathering. **Oxidation** (ahk sih DAY shun) occurs when some materials are exposed to oxygen and water. For example, when minerals containing iron are exposed to water and the oxygen in air, the iron in the min-eral reacts to form a new material that resembles rust. One com-mon example of this type of weathering is the alteration of the iron-bearing mineral magnetite to a rustlike material called limonite, as shown in **Figure 7.** Oxidation of minerals gives some rock layers a red color.

Reading Check *How does oxygen cause weathering?*

Effects of Climate

Climate affects soil temperature and moisture and also affects the rate of mechanical and chemical weathering. **Climate** is the pattern of weather that occurs in a particular area over many years. In cold climates, where freezing and thawing are frequent, mechanical weathering rapidly breaks down rock through the process of ice wedging. Chemical weathering is more rapid in warm, wet climates. High temperatures tend to increase the rate of chemical reactions. Thus, chemical weathering tends to occur quickly in tropical areas. Lack of moisture in deserts and low temperatures in polar regions slow down chemical weathering.

Magnetite

Limonite

Marble statue

Granite statue

Figure 8 Different types of rock weather at different rates. In humid climates, marble statues weather rapidly and become discolored. Granite statues weather more slowly.

Effects of Rock Type Rock type also can affect the rate of weathering in a particular climate. In wet climates, for example, marble weathers more rapidly than granite, as shown in **Figure 8.**

The weathering of rocks and the process of soil formation alter rock minerals so that soil minerals are mostly inherited from the parent rock type. Weathering begins the process of forming soil from rock and sediment and also affects particle size and soil texture. Recall that sand, silt, and clay simply describe the different particle sizes of the soil's mineral content.

section 1 review

Summary

Weathering and Its Effects

- Weathering includes processes that break down rock.
- Weathering affects Earth's landforms.

Mechanical Weathering

- During mechanical weathering, rock is broken apart, but it is not changed chemically.
- Plant roots, burrowing animals, and expanding ice all weather rock.

Chemical Weathering

- During chemical weathering, minerals in rock dissolve or change to other minerals.
- Agents of chemical weathering include natural acids and oxygen.

Self Check

1. **Describe** how weathering reduces the height of mountains through millions of years.
2. **Explain** how both tree roots and prairie dogs mechanically weather rock.
3. **Summarize** the effects of carbonic acid on limestone.
4. **Describe** how climate affects weathering.
5. **Think Critically** Why does limestone often form cliffs in dry climates but rarely form cliffs in wet climates?

Applying Skills

6. **Venn Diagram** Make a Venn diagram to compare and contrast mechanical weathering and chemical weathering. Include the causes of mechanical and chemical weathering in your diagram.

The Nature of Soil

as you read

What You'll Learn

- **Explain** how soil forms.
- **Describe** soil characteristics.
- **Describe** factors that affect the development of soil.

Why It's Important

Much of the food that you eat is grown in soil.

✏ Review Vocabulary

profile: a vertical slice through rock, sediment, or soil

New Vocabulary

- soil
- humus
- horizon
- soil profile
- litter
- leaching

Formation of Soil

The word *ped* is from a Greek word that means "ground" and from a Latin word that means "foot." The pedal under your foot, when you're bicycling, is named from the word *ped*. The part of Earth under your feet, when you're walking on the ground, is the pedosphere, or soil. Soil science is called pedology.

What is soil and where does it come from? A layer of rock and mineral fragments produced by weathering covers the surface of Earth. As you learned in Section 1, weathering gradually breaks rocks into smaller and smaller fragments. However, these fragments do not become high-quality soil until plants and animals live in them. Plants and animals add organic matter, the remains of once-living organisms, to the rock fragments. Organic matter can include leaves, twigs, roots, and dead worms and insects. **Soil** is a mixture of weathered rock, decayed organic matter, mineral fragments, water, and air.

Soil can take thousands of years to form and ranges from 60 m thick in some areas to just a few centimeters thick in others. Climate, slope, types of rock, types of vegetation, and length of time that rock has been weathering all affect the formation of soil, as shown in **Figure 9.** For example, different kinds of soils develop in tropical regions than in polar regions. Soils that develop on steep slopes are different from soils that develop on flat land. **Figure 10** illustrates how soil develops from rock.

Figure 9 Five different factors affect soil formation.

Explain *how time influences the development of soils.*

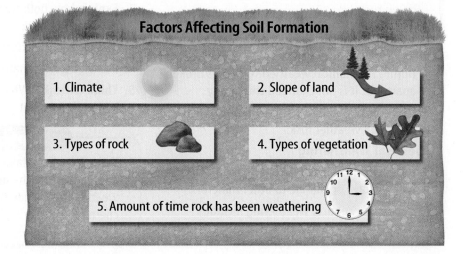

Factors Affecting Soil Formation

1. Climate
2. Slope of land
3. Types of rock
4. Types of vegetation
5. Amount of time rock has been weathering

Figure 10

It may take thousands of years to form, but soil is constantly evolving from solid rock, as this series of illustrations shows. Soil is a mixture of weathered rock, mineral fragments, and organic material—the remains of dead plants and animals—along with water and air.

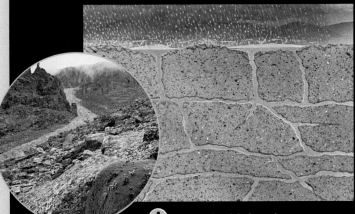

A Natural acids in rainwater weather the surface of exposed bedrock. Water can also freeze in cracks, causing rocks to fracture and break apart. The inset photo shows weathered rock in the Tien Shan Mountains of Central Asia.

B Plants take root in the cracks and among bits of weathered rock—shown in the inset photo above. As they grow, plants, along with other natural forces, continue the process of breaking down rocks, and a thin layer of soil begins to form.

C Like the grub in the inset photo, insects, worms, and other living things take up residence among plant roots. Their wastes, along with dead plant material, add organic matter to the soil.

D As organic matter increases and underlying bedrock continues to break down, the soil layer thickens. Rich topsoil supports trees and other plants with large root systems.

Mini LAB

Comparing Components of Soil

Procedure
1. Complete a safety worksheet.
2. Collect a sample of **soil.**
3. Observe it closely with a **magnifying lens.**
4. Record evidence of plant and animal components and their activities.

Analysis
1. Describe the different particles found in your sample. Did you find any remains of organisms?
2. Explain how living organisms might affect the soil.
3. Compare and contrast your sample with those other students have collected.

Composition of Soil

Soil is made up of rock and mineral fragments, organic matter, air, and water. The rock and mineral fragments come from rocks that have been weathered. Most of these fragments are small particles of sediment such as clay, silt, and sand.

Most organic matter in soil comes from plants. Plant leaves, stems, and roots all contribute organic matter to soil. Animals and microorganisms provide additional organic matter when they die. After plant and animal material gets into soil, fungi and bacteria cause it to decay. The decayed organic matter turns into a dark-colored material called **humus** (HYEW mus). Humus serves as a source of nutrients for plants. As worms, insects, and rodents burrow throughout soil, they mix the humus with the fragments of rock. Good-quality surface soil has approximately equal amounts of humus and weathered rock material.

Water Infiltration Soil has many small spaces between individual soil particles that are filled with water or air. When soil is moist, the spaces hold the water that plants need to grow. During a drought, the spaces are almost entirely filled with air. When water soaks into the ground, it infiltrates the pores. Infiltration rate is determined by calculating the time it takes for water sitting on soil to drop a fixed distance. This rate changes as the soil pore spaces fill with water.

Soil Profile

You have seen layers of soil if you've ever dug a deep hole or driven along a road that has been cut into a hillside. You probably observed that most plant roots grow in the top layer of soil. The top layer typically is darker than the soil layers below it. These different layers of soil are called **horizons.** All the horizons of a soil form a **soil profile.** Most soils have three horizons—labeled A, B, and C, as shown in **Figure 11.**

Figure 11 This soil, which developed beneath a grassy prairie, has three main horizons.
Describe *how the A horizon is different from the other two horizons.*

A horizon

B horizon

C horizon

A Horizon The A horizon is the top layer of soil. In a forest, the A horizon might be covered with litter. **Litter** consists of leaves, twigs, and other organic material that can be changed to humus by decomposing organisms. Litter helps prevent erosion and evaporation of water from soil. The A horizon also is known as topsoil. Topsoil has more humus and fewer rock and mineral particles than the other layers in a soil profile. The A horizon generally is dark and fertile. The dark color of the soil is caused by the humus, which provides nutrients for plant growth.

Since dark color absorbs solar energy more readily, soil color can greatly affect soil temperature. Darker color also may indicate a higher content of soil moisture. Soil moisture and soil temperature are important in determining seed germination for plants and the vitality of decomposing organisms.

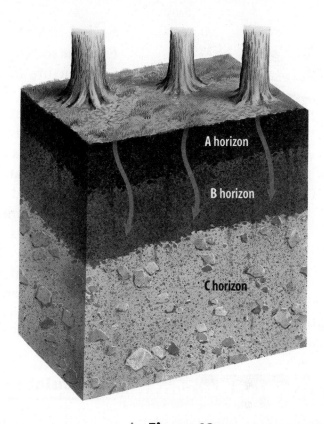

B Horizon The layer below the A horizon is the B horizon, also known as subsoil. Because less organic matter is added to this horizon, it is lighter in color than the A horizon and contains less humus. As a result, the B horizon is less fertile. The B horizon contains material moved down from the A horizon by the process of leaching.

Leaching is the removal of minerals that have been dissolved in water. The process of leaching resembles making coffee in a drip coffeemaker. In soil, water seeps through the A horizon and reacts with humus and carbon dioxide to form acid. The acid dissolves some of the minerals in the A horizon and carries the material into the B horizon, as shown in **Figure 12.**

Reading Check *How does leaching transport material from the A horizon to the B horizon?*

C Horizon The C horizon consists of partially weathered rock and is the bottom horizon in a soil profile. It is often the thickest soil horizon. This horizon does not contain much organic matter and is not strongly affected by leaching. It usually is composed of coarser sediment than the soil horizons above it. What would you find if you dug to the bottom of the C horizon? As you might have guessed, you would find rock—the rock that gave rise to the soil horizons above it. This rock is called the parent material of the soil. The C horizon is the soil layer that is most like the parent material.

Figure 12 Leaching removes material from the upper layer of soil. Much of this material then is deposited in the B horizon.

Soil Fertility Plants need a variety of nutrients for growth. They need things like nitrogen, phosphorus, potassium, sulfur, calcium, and magnesium called macronutrients They get these nutrients from the minerals and organic material in soil. Fertile soil supplies the nutrients that plants need in the proper amounts. Soil fertility usually is determined in a laboratory by a soil chemist. However, fertility sometimes can be inferred by looking at plants. Do research to discover more important plant nutrients.

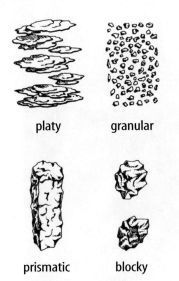

Figure 13 Four major classes characterize soil structure.

platy   granular

prismatic   blocky

Soil Structure Individual soil particles clump together. Examine soil closely and you will see natural clumps called *peds*. Soil structure affects pore space and will affect a plant's ability to penetrate roots. **Figure 13** shows four classes of soil structure. Granular structures are common in surface soils with high organic content that glues minerals together. Earthworms, frost, and rodents mix the soil, keeping the peds small, which provides good porosity and movement of air and water. Platy structures are often found in subsurface soils that have been leached or compacted by animals or machinery. Blocky structures are common in subsoils or surface soils with high clay content, which shrinks and swells, producing cracks. Prismatic structures, found in B horizons, are very dense and difficult for plant roots to penetrate. Vertical cracks result from freezing and thawing, wetting and drying, and downward movement of water and roots. Soil consistency refers to the ability of peds and soil particles to stick together and hold their shapes.

Applying Math   Calculate Percentages

SOIL TEXTURE Some soil is coarse, some is fine. This property of soil is called soil texture. The texture of soil often is determined by finding the percentages of sand, silt, and clay. Calculate the percentage of clay shown by the circle graph.

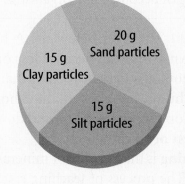

20 g
Sand particles

15 g
Clay particles

15 g
Silt particles

Solution

1 *This is what you know:*
- sand weight: 20 g
- clay weight: 15 g
- silt weight: 15 g

2 *This is what you need to find:*
- total weight of the sample
- percentage of clay particles

3 *This is the procedure you need to use:*
- Add all the masses to determine the total sample mass: 20 g sand + 15 g silt + 15 g clay = 50 g sample
- Divide the clay mass by the sample mass; multiply by 100: 15 g clay/50 g sample × 100 = 30% clay in the sample

Practice Problems

1. Calculate the percentage of sand in the sample.

2. Calculate the percentage of silt in the sample.

For more practice, visit bookg.msscience.com/ math_practice

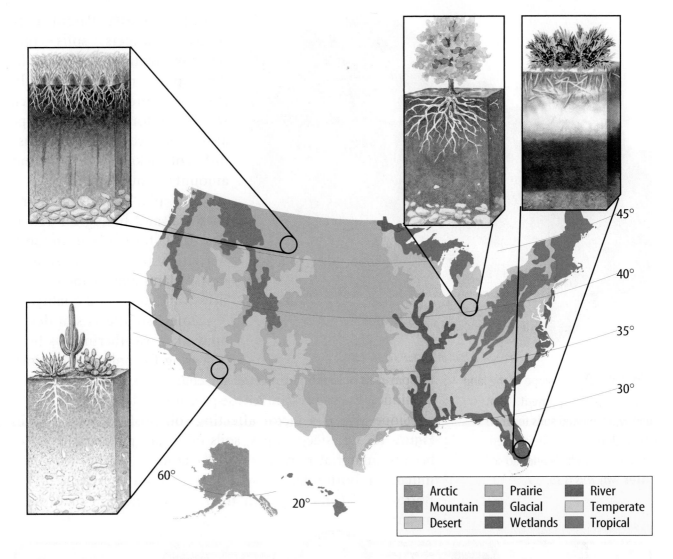

45°
40°
35°
30°

60°
20°

▨ Arctic	▨ Prairie	▨ River
▨ Mountain	▨ Glacial	▨ Temperate
▨ Desert	▨ Wetlands	▨ Tropical

Soil Types

If you travel across the country, you will notice that not all soils are the same. Some are thick and red. Some are brown with hard rock nodules, and some have thick, black A horizons. They vary in color, depth, texture, fertility, pH, temperature, and moisture content. Many soils exist, as shown in **Figure 14.**

Soil Types Reflect Climate Different regions on Earth have different climates. Deserts are dry, prairies are semidry, and temperate forests are mild and moist. These places also have different types of soils. Soil temperature and moisture content affect the quality of soils. Soils in deserts contain little organic material and also are thinner than soils in wetter climates. Prairie soils have thick, dark A horizons because the grasses that grow there contribute lots of organic matter. Temperate forest soils have less organic matter and thinner A horizons than prairie soils do. Other regions such as tundra and tropical areas also have distinct soils.

Figure 14 The United States has many different soil types. They vary in color, depth, texture, and fertility.
Identify *the soil type in your region.*

Other Factors Parent rock material affects soils that develop from it. Clay soils develop on rocks like basalt, because minerals in the rock weather to form clay. Rock type also affects vegetation, because different rocks provide different amounts of nutrients.

Soil pH, controls many chemical and biological activities that take place in soil. Activities of organisms, acid rain, or land management practices could affect soil quality.

Time also affects soil development. If weathering has been occurring for only a short time, the parent rock determines the soil characteristics. As weathering continues, the soil resembles the parent rock less and less.

Slope also is a factor affecting soil profiles, as shown in **Figure 15.** On steep slopes, soils often are poorly developed, because material moves downhill before it can be weathered much. In bottomlands, sediment and water are plentiful. Bottomland soils are often thick, dark, and full of organic material.

Figure 15 The slope of the land affects soil development. Thin, poorly developed soils form on steep slopes, but valleys often have thick, well-developed soils. **Infer** *why this is so.*

section 2 review

Summary

Formation of Soil
- Soil is a mixture of rock and mineral fragments, decayed organic matter, water, and air.

Composition of Soil
- Organic matter gradually changes to humus.
- Soil moisture is important for plant growth.

Soil Profile
- The layers in a soil profile are called horizons.
- Most soils have an A, B, and C horizon.

Soil Types
- Many different types of soils occur in the United States.
- Climate and other factors determine the type of soil that develops.

Self Check

1. **List** the five factors that affect soil development.
2. **Explain** how soil forms.
3. **Explain** why A horizons often are darker than B horizons or C horizons.
4. **Describe** how leaching affects soil.
5. **Think Critically** Why is a soil profile in a tropical rain forest different from one in a desert? A prairie?

Applying Skills

6. **Use Statistics** A farmer collected five soil samples from a field and tested their acidity, or pH. His data were the following: 7.5, 8.2, 7.7, 8.1, and 8.0. Calculate the mean of these data. Also, determine the range and median.

Soil Texture

Soils have different amounts of different sizes of particles. When you determine how much sand, silt, and clay a soil contains, you describe the soil's texture.

◉ Real-World Question

What is the texture of your soil?

Goals
■ **Estimate** soil texture by making a ribbon.

Materials

soil sample (100 g)   water bottle

Safety Precautions

◉ Procedure

1. Take some soil and make it into a ball. Work the soil with your fingers. Slowly add water to the soil until it is moist.

2. After your ball of soil is moist, try to form a thin ribbon of soil. Use the following descriptions to categorize your soil:
   a. If you can form a long, thin ribbon, you have a clay soil.
   b. If you formed a long ribbon but it breaks easily, you have a clay loam soil.
   c. If you had difficulty forming a long ribbon, you have loam soil.

3. Now make your soil classification more detailed by selecting one of these descriptions:
   a. If the soil feels smooth, add the word *silty* to your soil name.
   b. If the soil feels slightly gritty, don't add any word to your soil name.
   c. If the soil feels very gritty, add the word *sandy* before your soil name.

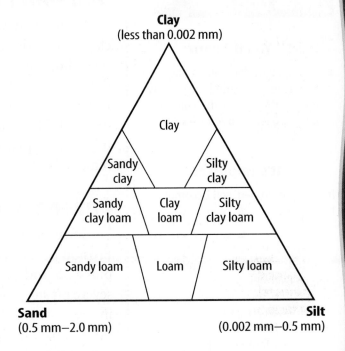

Clay
(less than 0.002 mm)

Clay

Sandy clay   Silty clay

Sandy clay loam   Clay loam   Silty clay loam

Sandy loam   Loam   Silty loam

Sand   **Silt**
(0.5 mm–2.0 mm)   (0.002 mm–0.5 mm)

◉ Conclude and Apply

1. **Classify** Which texture class name did you assign to your soil?

2. **Observe** Find your soil texture class name on the triangle above. Notice that the corners of the triangle are labeled *sand, silt,* and *clay.*

3. **Determine** Is your soil texture class close to one of the three corners or near the middle of the diagram? If your soil texture class is close to a corner, which one?

4. **Describe** Does your soil contain mostly sand, silt, or clay, or does it have nearly equal amounts of each? *Hint: If your soil name is close to a corner, it has mostly that size of sediment. If your soil name is in the middle of the triangle, it has nearly equal amounts of each sediment size.*

Soil Erosion

as you read

What You'll Learn

- **Explain** why soil is important.
- **Evaluate** ways that human activity has affected Earth's soil.
- **Describe** ways to reduce soil erosion.

Why It's Important

If topsoil is eroded, soil becomes less fertile.

⊙ Review Vocabulary

erosion: the picking up and moving of sediment or soil

New Vocabulary

- no-till farming
- contour farming
- terracing

Figure 16 Removing vegetation can increase soil erosion.

Soil—An Important Resource

While picnicking at a local park, a flash of lightning and a clap of thunder tell you that a storm is upon you. Watching the pounding rain from the park shelter, you notice that the water flowing off of the ball diamond is muddy, not clear. The flowing water is carrying away some of the sediment that used to be on the field. This process is called soil erosion. Soil erosion is harmful because plants do not grow as well when topsoil has been removed.

Causes and Effects of Soil Erosion

Soil erodes when it is moved from the place where it formed. Erosion occurs as water flows over Earth's surface or when wind picks up and transports sediment. Generally, erosion is more severe on steep slopes than on gentle slopes. It's also more severe in areas where there is little vegetation. Under normal conditions, a balance between soil production and soil erosion often is maintained. This means that soil forms at about the same rate as it erodes. However, humans sometimes cause erosion to occur faster than new soil can form. One example is when people remove ground cover. Ground cover is vegetation that covers the soil and protects it from erosion. When vegetation is cleared, as shown in **Figure 16,** soil erosion often increases.

Trees protect the soil from erosion in forested regions.

When forest is removed, soil erodes rapidly.

Figure 17 Tropical rain forests often are cleared by burning. **Explain** *how this can increase soil erosion.*

Agricultural Cultivation

Soil erosion is a serious problem for agriculture. Topsoil contains many nutrients, holds water well, and has a porous structure that is good for plant growth. If topsoil is eroded, the quality of the soil is reduced. For example, plants need nutrients to grow. Each year, nutrients are both added to the soil and removed from the soil. The difference between the amount of nutrients added and the amount of nutrients removed is called the nutrient balance. If topsoil erodes rapidly, the nutrient balance might be negative. Farmers might have to use more fertilizer to compensate for the nutrient loss. In addition, the remaining soil might not have the same open structure and water-holding ability that topsoil does.

Forest Harvesting

When forests are removed, soil is exposed and erosion increases. This creates severe problems in many parts of the world, but tropical regions are especially at risk. Each year, thousands of square kilometers of tropical rain forest are cleared for lumber, farming, and grazing, as shown in **Figure 17.** Soils in tropical rain forests appear rich in nutrients but are almost infertile below the first few centimeters. The soil is useful to farmers for only a few years before the topsoil is gone. Farmers then clear new land, repeating the process and increasing the damage to the soil.

Overgrazing

In most places, land can be grazed with little damage to soil. However, overgrazing can increase soil erosion. In some arid regions of the world, sheep and cattle raised for food are grazed on grasses until almost no ground cover remains to protect the soil. When natural vegetation is removed from land that receives little rain, plants are slow to grow back. Without protection, soil is carried away by wind, and the moisture in the soil evaporates.

INTEGRATE Career

Soil Scientist Elvia Niebla is a soil scientist at the U.S. Environmental Protection Agency (EPA). Soil scientists at the EPA work to reduce soil erosion and pollution. Niebla's research even helped keep hamburgers safe to eat. How? In a report for the EPA, she explained how meat can be contaminated when cattle graze on polluted soil.

Science Online

Topic: Land Use
Visit bookg.msscience.com for Web links to information about how land use affects Earth's soil and about measures taken to reduce the impact.

Activity Debate with classmates about the best ways to protect rich farmland. Consider advantages and disadvantages of each method.

Excess Sediment If soil erosion is severe, sediment can damage the environment. Severe erosion sometimes occurs where land is exposed. Examples might include strip-mined areas or large construction sites. Eroded soil is moved to a new location where it is deposited. If the sediment is deposited in a stream, as shown in **Figure 18,** the stream channel might fill.

Figure 18 Erosion from exposed land can cause streams to fill with excessive amounts of sediment. **Explain** *how this could damage streams.*

Preventing Soil Erosion

Each year more than 1.5 billion metric tons of soil are eroded in the United States. Soil is a natural resource that must be managed and protected. People can do several things to conserve soil.

Manage Crops All over the world, farmers work to slow soil erosion. They plant shelter belts of trees to break the force of the wind and plant crops to cover the ground after the main harvest. In dry areas, instead of plowing under crops, many farmers graze animals on the vegetation. Proper grazing management can maintain vegetation and reduce soil erosion.

In recent years, many farmers have begun to practice no-till farming. Normally, farmers till or plow their fields one or more times each year. Using **no-till farming,** seen in

Figure 19 No-till farming decreases soil erosion because fields are not plowed.

Figure 19, farmers leave plant stalks in the field over the winter months. At the next planting, they seed crops without destroying these stalks and without plowing the soil. Farm machinery makes a narrow slot in the soil, and the seed is planted in this slot. No-till farming provides cover for the soil year-round, which reduces water runoff and soil erosion. One study showed that no-till farming can leave as much as 80% of the soil covered by plant residue. The leftover stalks also keep weeds from growing in the fields.

✓ **Reading Check** *How can farmers reduce soil erosion?*

Reduce Erosion on Slopes On gentle slopes, planting along the natural contours of the land, called **contour farming,** reduces soil erosion. This practice, shown in **Figure 20,** slows the flow of water down the slope and helps prevent the formation of gullies.

Where slopes are steep, terracing often is used. **Terracing** (TER uh sing) is a method in which steep-sided, level topped areas are built onto the sides of steep hills and mountains so that crops can be grown. These terraces reduce runoff by creating flat areas and shorter sections of slope. In the Philippines, Japan, China, and Peru, terraces have been used for centuries.

Figure 20 This orchard was planted along the natural contours of the land.
Summarize *the benefits of using contour farming on slopes.*

Reduce Erosion of Exposed Soil A variety of methods are used to control erosion where soil is exposed. During the construction process water is sometimes sprayed onto bare soil to prevent erosion by wind. When construction is complete, topsoil is added in areas where it was removed and trees are planted. At strip mines, water flow can be controlled so that most of the eroded soil is kept from leaving the mine. After mining is complete, the land is reclaimed. This means that steep slopes are flattened and vegetation is planted.

section 3 review

Summary

Soil—An Important Resource

- Soil erosion is a serious problem because topsoil is removed from the land.

Causes and Effects of Soil Erosion

- Soil erosion occurs rapidly on steep slopes and areas that are not covered by vegetation.
- The quality of farmland is reduced when soil erosion occurs.

Preventing Soil Erosion

- Farmers reduce erosion by planting shelter belts, using no-till farming, and planting cover crops after harvesting.
- Contour farming and terracing are used to control erosion on slopes.

Self Check

1. **Explain** why soil is important.
2. **Explain** how soil erosion damages soil.
3. **Describe** no-till farming.
4. **Explain** how overgrazing increases soil erosion.
5. **Think Critically** How does contour farming help water soak into the ground?

Applying Skills

6. **Communicate** Do research to learn about the different methods that builders use to reduce soil erosion during construction. Write a newspaper article describing how soil erosion at large construction sites is being controlled in your area.

Design Your Own

WEATHERING CHALK

Goals

- **Design** experiments to evaluate the effects of acidity, surface area, and temperature on the rate of chemical weathering of chalk.
- **Describe** factors that affect chemical weathering.
- **Explain** how the chemical weathering of chalk is similar to the chemical weathering of rocks.

Possible Materials

pieces of chalk (6)
small beakers (2)
metric ruler
water
white vinegar (100 mL)
hot plate
computer probe for temperature
*thermometer
*Alternate materials

Safety Precautions

Wear safety goggles when pouring vinegar. Be careful when using a hot plate and heated solutions.

WARNING: *If mixing liquids, always add acid to water.*

◉ Real-World Question

Chalk is a type of limestone made of the shells of microscopic organisms. The famous White Cliffs of Dover, England, are made up of chalk. This lab will help you understand how chalk can be chemically weathered. How can you simulate chemical weathering of chalk?

◉ Form a Hypothesis

How do you think acidity, surface area, and temperature affect the rate of chemical weathering of chalk? What happens to chalk in water? What happens to chalk in acid (vinegar)? How will the size of the chalk pieces affect the rate of weathering? What will happen if you heat the acid? Make hypotheses to support your ideas.

Test Your Hypothesis

Make a Plan

1. **Develop** hypotheses about the effects of acidity, surface area, and temperature on the rate of chemical weathering.

2. **Decide** how to test your first hypothesis. List the steps needed to test the hypothesis.

3. Repeat step 2 for your other two hypotheses.

4. **Design** data tables in your Science Journal. Make one for acidity, one for surface area, and one for temperature.

5. **Identify** what remains constant in your experiment and what varies. Change only one variable in each procedure.

6. **Summarize** your data in a graph. Decide from reading the Science Skill Handbook which type of graph to use.

Follow Your Plan

1. Make sure your teacher approves your plan before you start.

2. Carry out the three experiments as planned.

3. While you are conducting the experiments, record your observations and complete the data tables in your Science Journal.

4. Graph your data to show how each variable affected the rate of weathering.

Analyze Your Data

1. **Analyze** your graph to find out which substance—water or acid—weathered the chalk more quickly. Was your hypothesis supported by your data?

2. **Infer** from your data whether the amount of surface area makes a difference in the rate of chemical weathering. Explain.

Conclude and Apply

1. **Explain** how the chalk was chemically weathered.

2. How does heat affect the rate of chemical weathering?

3. What does this imply about weathering in the tropics and in polar regions?

ommunicating
Your Data

Compare your results with those of your classmates. How were your data similar? How were they different? **For more help, refer to the** Science Skill Handbook.

Landscape, History, and the Pueblo Imagination

by Leslie Marmon Silko

Leslie Marmon Silko, a woman of Pueblo, Hispanic, and American heritage, explains what ancient Pueblo people believed about the circle of life on Earth.

You see that after a thing is dead, it dries up. It might take weeks or years, but eventually if you touch the thing, it crumbles under your fingers. It goes back to dust. The soul of the thing has long since departed. With the plants and wild game the soul may have already been borne back into bones and blood or thick green stalk and leaves. Nothing is wasted. What cannot be eaten by people or in some way used must then be left where other living creatures may benefit. What domestic animals or wild scavengers can't eat will be fed to the plants. The plants feed on the dust of these few remains.

. . . Corn cobs and husks, the rinds and stalks and animal bones were not regarded by the ancient people as filth or garbage. The remains were merely resting at a mid-point in their journey back to dust. . . .

The dead become dust . . . . The ancient Pueblo people called the earth the Mother Creator of all things in this world. Her sister, the Corn mother, occasionally merges with her because all . . . green life rises out of the depths of the earth.

Rocks and clay . . . become what they once were. Dust.

A rock shares this fate with us and with animals and plants as well.

Understanding Literature

Repetition The recurrence of sounds, words, or phrases is called repetition. What is Silko's purpose of the repeated use of the word *dust?*

Respond to the Reading

1. What one word is repeated throughout this passage?
2. What effect does the repetition of this word have on the reader?
3. **Linking Science and Writing** Using repetition, write a one-page paper on how to practice a type of soil conservation.

INTEGRATE Earth Science This chapter discusses how weathered rocks and mineral fragments combine with organic matter to make soil. Silko's writing explains how the ancient Pueblo people understood that all living matter returns to the earth, or becomes dust. Lines such as "green life rises out of the depths of the earth," show that the Pueblo people understood that the earth, or rocks and mineral fragments, must combine with living matter in order to make soil and support plant life.

Reviewing Main Ideas

Section 1 Weathering

1. Weathering helps to shape Earth's surface.

2. Mechanical weathering breaks apart rock without changing its chemical composition. Plant roots, animals, and ice wedging are agents of mechanical weathering.

3. Chemical weathering changes the chemical composition of rocks. Natural acids and oxygen in the air can cause chemical weathering.

Section 2 The Nature of Soil

1. Soil is a mixture of rock and mineral fragments, organic matter, air, and water.

2. A soil profile contains different layers that are called horizons.

3. Climate, parent rock, slope of the land, type of vegetation, and the time that rock has been weathering are factors that affect the development of soil.

Section 3 Soil Erosion

1. Soil is eroded when it is moved to a new location by wind or water.

2. Human activities can increase the rate of soil erosion.

3. Windbreaks, no-till farming, contour farming, and terracing reduce soil erosion on farm fields.

Visualizing Main Ideas

Copy and complete the following concept map about weathering.

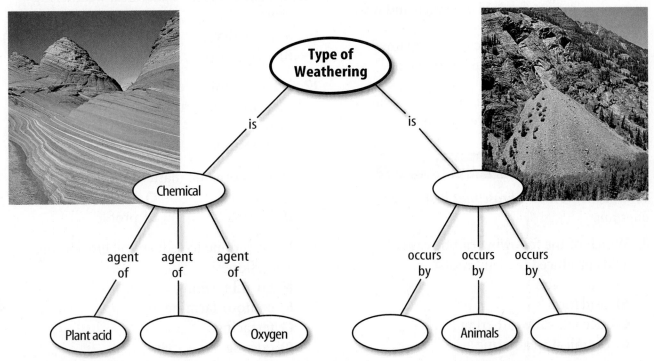

Using Vocabulary

chemical weathering p. 39	mechanical weathering p. 37
climate p. 40	no-till farming p. 52
contour farming p. 53	oxidation p. 40
horizon p. 44	soil p. 42
humus p. 44	soil profile p. 44
ice wedging p. 38	terracing p. 53
leaching p. 45	weathering p. 36
litter p. 45	

Fill in the blanks with the correct vocabulary word or words.

1. _____ changes the composition of rock.

2. _____ forms from organic matter such as leaves and roots.

3. The horizons of a soil make up the _____.

4. _____ transports material to the B horizon.

5. _____ occurs when many materials containing iron are exposed to oxygen and water.

6. _____ means that crops are planted along the natural contours of the land.

7. _____ is the pattern of weather that occurs in a particular area for many years.

Checking Concepts

Choose the word or phrase that best answers the question.

8. Which of the following can be caused by acids produced by plant roots?
   A) soil erosion
   B) oxidation
   C) mechanical weathering
   D) chemical weathering

Use the graph below to answer question 9.

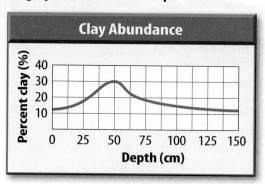

9. The above graph shows the percentage of clay in a soil profile at varying depths. Which depth has the highest amount of clay?
   A) 25 cm
   B) 150 cm
   C) 50 cm
   D) 100 cm

10. Which of the following is an agent of mechanical weathering?
    A) animal burrowing
    B) carbonic acid
    C) leaching
    D) oxidation

11. In which region is chemical weathering most rapid?
    A) cold, dry
    B) cold, moist
    C) warm, moist
    D) warm, dry

12. What is a mixture of rock and mineral fragments, organic matter, air, and water called?
    A) soil
    B) limestone
    C) horizon
    D) clay

13. What is organic matter in soil?
    A) leaching
    B) humus
    C) horizon
    D) profile

14. What is done to reduce soil erosion on steep slopes?
    A) no-till farming
    B) contour farming
    C) terracing
    D) grazing

Science Online bookg.msscience.com/vocabulary_puzzlemaker

Thinking Critically

15. Predict which type of weathering—mechanical or chemical—you would expect to have a greater effect in a polar region. Explain.

16. Recognize Cause and Effect How does soil erosion reduce the quality of soil?

17. Concept Map Copy and complete the concept map about layers in soil.

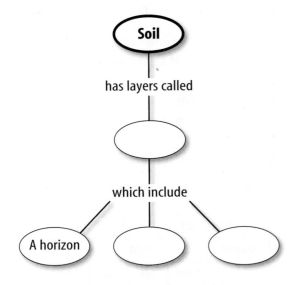

18. Recognize Cause and Effect Why do rows of trees along the edges of farm fields reduce wind erosion of soil?

19. Form a Hypothesis A pile of boulders lies at the base of a high-mountain cliff. Form a hypothesis explaining how the pile of rock might have formed.

20. Test a Hypothesis How would you test your hypothesis from question 19?

21. Identify a Question Many scientists are conducting research to learn more about how soil erosion occurs and how it can be reduced. Write a question about soil erosion that you would like to research. With your teacher's help, carry out an investigation to answer your question.

Performance Activities

22. Design a Landscape Find a slope in your area that might benefit from erosion maintenance. Develop a plan for reducing erosion on this slope. Make a map showing your plan.

23. Describing Peds Natural clumps of soil are called peds. Collect a large sample of topsoil. Describe the shape of the peds. Sketch the peds in your Science Journal.

Applying Math

Use the illustration below to answer questions 24–26.

24. Fertilizer Nutrients A bag of fertilizer is labeled to list the nutrients as three numbers. The numbers represent the percentages of nitrogen, phosphate, and potash in that order. What are the percentages of these nutrients for a fertilizer with the following information on the label: 5-10-10?

25. Fertilizer Ratio The fertilizer ratio tells you the proportions of the different nutrients in a fertilizer. To find the fertilizer ratio, divide each nutrient value by the lowest value. Calculate the fertilizer ratio for the fertilizer in question 24.

26. Relative Amounts of Nutrients Which nutrient is least abundant in the fertilizer? Which nutrients are most abundant? How many times more potash does the fertilizer contain than nitrogen?

Part 1 | Multiple Choice

Record your answers on the answer sheet provided by your teacher or on a separate sheet of paper.

Use the photo below to answer question 1.

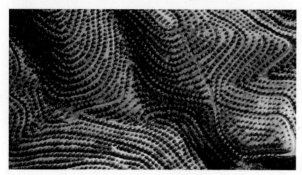

1. Which method for reducing soil erosion is shown on the hillsides above?
   A. no-till farming   C. contour farming
   B. terracing   D. shelter belts

2. Which of the following terms might describe a soil's texture?
   A. red   C. porous
   B. coarse   D. wet

3. Which soil horizon often has a dark color because of the presence of humus?
   A. E horizon   C. B horizon
   B. C horizon   D. A horizon

4. Which of the following is an agent of chemical weathering?
   A. ice wedging
   B. burrowing animals
   C. carbonic acid
   D. growing tree roots

Test-Taking Tip

Come Back To It Never skip a question. If you are unsure of an answer, mark your best guess on another sheet of paper and mark the question in your test booklet to remind you to come back to it at the end of the test.

5. Which of the following might damage a soil's structure?
   A. a gentle rain   C. earthworms
   B. organic matter   D. compaction

6. In which of the following types of rock are caves most likely to form?
   A. limestone   C. granite
   B. sandstone   D. basalt

7. Which of the following is most likely to cause erosion of farmland during a severe drought?
   A. water runoff
   B. soil creeping downhill
   C. wind
   D. ice

Use the table below to answer questions 8–10.

Texture Data for a Soil Profile			
Horizon	Percent		
	Sand	Silt	Clay
A	16.2	54.4	29.4
B	10.5	50.2	39.3
C	31.4	48.4	20.2
R (bedrock)	31.7	50.1	18.2

8. According to the table, which horizon in this soil has the lowest percentage of sand?
   A. A horizon   C. C horizon
   B. B horizon   D. R horizon

9. Which of the following is the R horizon?
   A. topsoil   C. bedrock
   B. humus   D. gravel

10. Which of the following is the best description of the soil represented by the table?
    A. sandy   C. clayey
    B. silty   D. organic

Part 2 | Short Response/Grid In

*Record your answers on the answer sheet
provided by your teacher or on a sheet of paper.*

Use the illustration below to answer question 11.

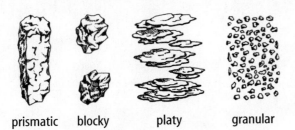

prismatic   blocky   platy   granular

11. Natural clumps of soil are called peds.
Compare and contrast the different types
of peds in the sketch. Explain how the
names of the peds describe their shape.

12. Explain how caves form. What role does
carbonic acid have?

Use the diagram below to answer questions 13–15.

2003 Phosphorus Budget per Acre
Jones Family Farm

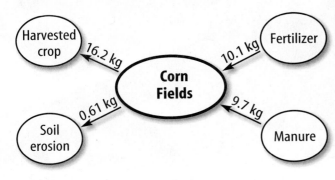

13. According to the diagram above, what was
the total amount of phosphorus added to
each acre?

14. What is the total amount of phorphorus
lost from each acre?

15. What is the difference between the
amount of phosphorus added and the
amount of phosphorus lost?

Part 3 | Open Ended

Record your answers on a sheet of paper.

16. Describe ways that humans affect Earth's
soil. How can damage to soil be reduced?

17. How does weathering change Earth's
surface?

18. How does no-till farming reduce soil
erosion?

19. How does time affect soil development?

20. How does humus form? What does it
form from?

Use the graph below to answer questions 21–23.

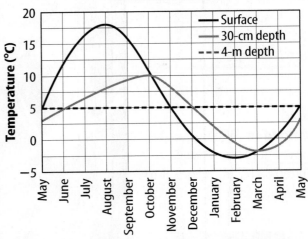

21. During which month was the surface soil
warmest? During which month was it
coldest? Explain.

22. During which month was the soil at 30 cm
warmest? During which month was it
coldest? How is this different than surface
soil?

23. Why didn't soil temperature vary at a
depth of 4 m?

Erosional Forces

For Sale—Ocean View

This home, once perched on a hillside overlooking the water, has been destroyed by landslides and flooding. In this chapter you will learn how large amounts of soil, such as the soil that once supported this house, can move from one place to another.

Science Journal Name three major landforms around the world and explain what erosional forces helped shape them.

Start-Up Activities

Demonstrate Sediment Movement

Can you think of ways to move something without touching it? In nature, sediment is moved from one location to another by a variety of forces. What are some of these forces? In this lab, you will investigate to find out the answers to these questions.

WARNING: *Do not pour sand or gravel down the drain.*

1. Place a small pile of a sand-and-gravel mixture into a large shoe-box lid.

2. Move the sediment pile to the other end of the lid without touching the particles with your hands. You can touch and manipulate the box lid.

3. Try to move the mixture in a number of different ways.

4. **Think Critically** In your Science Journal, describe the methods you used to move the sediment. Which method was most effective? Explain how your methods compare with forces of nature that move sediment.

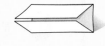

Erosion and Deposition Make the following Foldable to help you identify the examples of erosion and deposition.

STEP 1 Fold one piece of paper widthwise into thirds.

STEP 2 Fold the paper lengthwise into fourths.

STEP 3 Unfold, lay the paper lengthwise, and draw lines along the folds.

STEP 4 Label your table as shown.

Erosional Force	Erosion	Deposition
Gravity		
Glaciers		
Wind		

Make a Table As you read the chapter, complete the table, listing specific examples of erosion and deposition for each erosional force.

Preview this chapter's content and activities at bookg.msscience.com

Identify Cause and Effect

① Learn It! A cause is the reason something happens. The result of what happens is called an effect. Learning to identify causes and effects helps you understand why things happen. By using graphic organizers, you can sort and analyze causes and effects as you read.

② Practice It! Read the following paragraph. Then use the graphic organizer below to show what happened when ice freezes in the cracks of rocks.

…Rockfalls happen when blocks of rock break loose from a steep slope and tumble through the air. As they fall, these rocks crash into other rocks and knock them loose. More and more rocks break loose and tumble to the bottom. The fall of a single, large rock down a steep slope can cause serious damage to structures at the bottom. During the winter, when ice freezes in the cracks of rocks, the cracks expand and extend. In the spring, the pieces of rock break loose and fall down the mountainside…

—from page 66

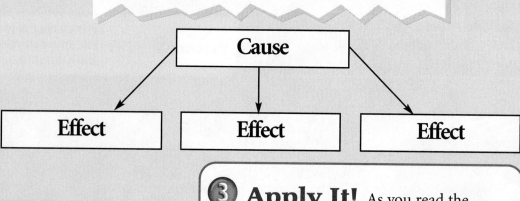

③ Apply It! As you read the chapter, be aware of causes and effects of gravity and ice. Find five causes and their effects.

Target Your Reading

Reading Tip

Graphic organizers such as the Cause-Effect organizer help you organize what you are reading so you can remember it later.

Use this to focus on the main ideas as you read the chapter.

1 **Before you read** the chapter, respond to the statements below on your worksheet or on a numbered sheet of paper.

- Write an **A** if you **agree** with the statement.
- Write a **D** if you **disagree** with the statement.

2 **After you read** the chapter, look back to this page to see if you've changed your mind about any of the statements.

- If any of your answers changed, explain why.
- Change any false statements into true statements.
- Use your revised statements as a study guide.

Science Online

Print out a worksheet of this page at bookg.msscience.com

Before You Read A or D		Statement	After You Read A or D
	1	Gravity, water, wind, and glaciers are common agents of erosion.	
	2	Deposition occurs when agents of erosion lose energy and drop the sediments they are carrying.	
	3	Mass movement is the slow process of changing rock into soil.	
	4	The two broad categories of glaciers are called continental glaciers and valley glaciers.	
	5	During the most recent ice age, continental glaciers covered the entire Earth.	
	6	Valley glaciers carve deep, V-shaped valleys.	
	7	Abrasion can be caused by windblown sediment striking and wearing away the surface of rock.	
	8	Most dunes move, or migrate away from the direction of the wind.	
	9	During sandstorms, large sand grains are often carried high into the atmosphere.	

Erosion by Gravity

as you read

What You'll Learn

- **Explain** the differences between erosion and deposition.
- **Compare** and contrast slumps, creep, rockfalls, rock slides, and mudflows.
- **Explain** why building on steep slopes might not be wise.

Why It's Important

Many natural features throughout the world were shaped by erosion.

Review Vocabulary

sediment: loose materials, such as mineral grains and rock fragments, that have been moved by erosional forces

New Vocabulary

- erosion
- deposition
- mass movement
- slump
- creep

Erosion and Deposition

Do you live in an area where landslides occur? As **Figure 1** shows, large piles of sediment and rock can move downhill with devastating results. Such events often are triggered by heavy rainfall. The muddy debris at the lower end of the slide comes from material that once was further up the hillside. The displaced soil and rock debris is a product of erosion (ih ROH zhun). **Erosion** is a process that wears away surface materials and moves them from one place to another.

What wears away sediments? How were you able to move the pile of sediments in the Launch Lab? If you happened to tilt the pan, you took advantage of an important erosional force—gravity. Gravity is the force of attraction that pulls all objects toward Earth's center. Other causes of erosion, also called agents of erosion, are water, wind, and glaciers.

Water and wind erode materials only when they have enough energy of motion to do work. For example, air can't move much sediment on a calm day, but a strong wind can move dust and even larger particles. Glacial erosion works differently by slowly moving sediment that is trapped in solid ice. As the ice melts, sediment is deposited, or dropped. Sometimes sediment is carried farther by moving meltwater.

Figure 1 The jumbled sediment at the base of a landslide is material that once was located farther uphill.
Define *the force that moves materials toward the center of Earth.*

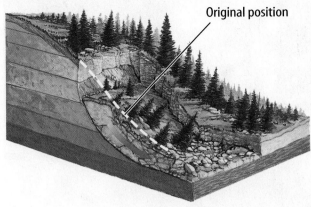

Original position

Dropping Sediments Agents of erosion drop the sediments they are carrying as they lose energy. This is called **deposition.** When sediments are eroded, they are not lost from Earth—they are just relocated.

Mass Movement

The greater an object's mass is, the greater its gravitational force is. Earth has such a great mass that gravity is a major force of erosion and deposition. Rocks and other materials, especially on steep slopes, are pulled toward the center of Earth by gravity.

A **mass movement** is any type of erosion that happens as gravity moves materials downslope. Some mass movements are so slow that you hardly notice they're happening. Others happen quickly—possibly causing catastrophes. Common types of mass movement include slump, creep, rockfalls, rock slides, and mudflows. Landslides are mass movements that can be one of these types or a combination of these types of mass movement.

Reading Check *What is a mass movement?*

Slump When a mass of material slips down along a curved surface, the mass movement is called **slump.** Often, when a slope becomes too steep, the base material no longer can support the rock and sediment above it. The soil and rock slip downslope as one large mass or break into several sections.

Sometimes a slump happens when water moves to the base of a slipping mass of sediment. This water weakens the slipping mass and can cause movement of material downhill. Or, if a strong rock layer lies on top of a weaker layer—commonly clay—the clay can weaken further under the weight of the rock. The clay no longer can support the strong rock on the hillside. As shown in **Figure 2,** a curved scar is left where the slumped materials originally rested.

Figure 2 Slump occurs when material slips downslope as one large mass.
Infer *What might have caused this slump to happen?*

Mini LAB

Modeling Slump
Procedure
WARNING: *Do not pour lab materials down the drain.*
1. Place one end of a **baking pan** on **two bricks** and position the other end over a sink with a sealed drain.
2. Fill the bottom half of the pan with **gelatin powder** and the top half of the pan with **aquarium gravel.** Place a large, **flat rock** on the gravel.
3. Using a **watering can,** sprinkle water on the materials in the pan for several minutes. Record your observations in your **Science Journal.**

Analysis
1. What happened to the different sediments in the pan?
2. Explain how your experiment models slump.

Figure 3 Over time, creep has caused these tree trunks to lean downhill. The trees then curved back toward the Sun.

Figure 4 Rockfalls, such as this one, occur as material free falls through the air.

Creep The next time you travel, look along the roadway or trail for slopes where trees and fence posts lean downhill. Leaning trees and human-built structures show another mass movement called creep. **Creep** occurs when sediments slowly shift their positions downhill, as **Figure 3** illustrates. Creep is common in areas of frequent freezing and thawing.

Rockfalls and Rock Slides Signs along mountainous roadways warn of another type of mass movement called rockfalls. Rockfalls happen when blocks of rock break loose from a steep slope and tumble through the air. As they fall, these rocks crash into other rocks and knock them loose. More and more rocks break loose and tumble to the bottom. The fall of a single, large rock down a steep slope can cause serious damage to structures at the bottom. During the winter, when ice freezes in the cracks of rocks, the cracks expand and extend. In the spring, the pieces of rock break loose and fall down the mountainside, as shown in **Figure 4.**

Rock slides occur when layers of rock—usually steep layers—slip downslope suddenly. Rock slides, like rockfalls, are fast and can be destructive in populated areas. They commonly occur in mountainous areas or in areas with steep cliffs, also as shown in **Figure 5.** Rock slides happen most often after heavy rains or during earthquakes, but they can happen on any rocky slope at any time without warning.

Figure 5 Rock slides are common in regions where layers of rock are steep.

Mudflows What would happen if you took a long trip and forgot to turn off the sprinkler in your hillside garden before you left? If the soil is usually dry, the sprinkler water could change your yard into a muddy mass of material much like chocolate pudding. Part of your garden might slide downhill. You would have made a mudflow, a thick mixture of sediments and water flowing down a slope. The mudflow in **Figure 6** caused a lot of destruction.

Mudflows usually occur in areas that have thick layers of loose sediments. They often happen after vegetation has been removed by fire. When heavy rains fall on these areas, water mixes with sediment, causing it to become thick and pasty. Gravity causes this mass to flow downhill. When a mudflow finally reaches the bottom of a slope, it loses its energy of motion and deposits all the sediment and everything else it has been carrying. These deposits often form a mass that spreads out in a fan shape. Why might mudflows cause more damage than floodwaters?

Figure 6 Mudflows, such as these in the town of Sarno, Italy, have enough energy to move almost anything in their paths. **Explain** *how mudflows differ from slumps, creep, and rock slides.*

 **Reading Check** *What conditions are favorable for triggering mudflows?*

Mudflows, rock slides, rockfalls, creep, and slump are similar in some ways. They all are most likely to occur on steep slopes, and they all depend on gravity to make them happen. Also, all types of mass movement occur more often after a heavy rain. The water adds mass and creates fluid pressure between grains and layers of sediment. This makes the sediment expand—possibly weakening it.

Consequences of Erosion

People like to have a great view and live in scenic areas away from noise and traffic. To live this way, they might build or move into houses and apartments on the sides of hills and mountains. When you consider gravity as an agent of erosion, do you think steep slopes are safe places to live?

Building on Steep Slopes When people build homes on steep slopes, they constantly must battle naturally occurring erosion. Sometimes builders or residents make a slope steeper or remove vegetation. This speeds up the erosion process and creates additional problems. Some steep slopes are prone to slumps because of weak sediment layers underneath.

INTEGRATE Physics

Driving Force The force that drives most types of erosion is gravity. Water at an elevation has potential, or stored energy. When water drops in elevation this energy changes to kinetic energy, or energy of motion. Water may then become a powerful agent of erosion. Find out how water has shaped the region in which you live.

Making Steep Slopes Safe Plants can be beautiful or weedlike—but they all have root structures that hold soil in place. One of the best ways to reduce erosion is to plant vegetation. Deep tree roots and fibrous grass roots bind soil together, reducing the risk of mass movement. Plants also absorb large amounts of water. Drainage pipes or tiles inserted into slopes can prevent water from building up, too. These materials help increase the stability of a slope by allowing excess water to flow out of a hillside more easily.

Walls made of concrete or boulders also can reduce erosion by holding soil in place, as shown in **Figure 7.** However, preventing mass movements on a slope is difficult because rain or earthquakes can weaken all types of Earth materials, eventually causing them to move downhill.

✔ Reading Check *What can be done to slow erosion on steep slopes?*

Figure 7 Some slopes are stabilized by building walls made from concrete or stone.

People who live in areas with erosion problems spend a lot of time and money trying to preserve their land. Sometimes they're successful in slowing down erosion, but they never can eliminate erosion and the danger of mass movement. Eventually, gravity wins. Sediment moves from place to place, constantly reducing elevation and changing the shape of the land.

---

section ① review

Summary

Erosion and Deposition

- Gravity is the force that pulls all objects toward Earth's center.
- Water and wind erode materials only when they have enough energy of motion to do work.
- Agents of erosion drop sediment as they lose energy.

Mass Movement

- The greater an object's mass is, the greater its gravitational force is.
- Gravity is a major force of erosion and deposition.
- Common types of mass movement include slump, creep, rockfalls, rock slides, and mudflows.

Self Check

1. **Define** the term *erosion* and name the forces that cause it.
2. **Explain** how deposition changes the surface of Earth.
3. **Describe** the characteristics that all types of mass movements have in common.
4. **Describe** ways to help slow erosion on steep slopes.
5. **Think Critically** When people build houses and roads, they often pile up dirt or cut into the sides of hills. Predict how this might affect sediment on a slope. Explain how to control the effects of such activities.

Applying Skills

6. **Compare and Contrast** What are the similarities and differences between rock falls and rock slides?

Glaciers

How Glaciers Form and Move

If you've ever gone sledding, snowboarding, or skiing, you might have noticed that after awhile, the snow starts to pack down under your weight. A snowy hillside can become icy if it is well traveled. In much the same way, glaciers form in regions where snow accumulates. Some areas of the world, as shown in **Figure 8,** are so cold that snow remains on the ground year-round. When snow doesn't melt, it piles up. As it accumulates slowly, the increasing weight of the snow becomes great enough to compress the lower layers into ice. Eventually, there can be enough pressure on the ice so that it becomes plasticlike. The mass slowly begins to flow in a thick, plasticlike lower layer, and ice slowly moves away from its source. A large mass of ice and snow moving on land under its own weight is a **glacier.**

Ice Eroding Rock

Glaciers are agents of erosion. As glaciers pass over land, they erode it, changing features on the surface. Glaciers then carry eroded material along and deposit it somewhere else. Glacial erosion and deposition change large areas of Earth's surface. How is it possible that something as fragile as snow or ice can push aside trees, drag rocks along, and slowly change the surface of Earth?

as you read

What You'll Learn

- **Explain** how glaciers move.
- **Describe** evidence of glacial erosion and deposition.
- **Compare and contrast** till and outwash.

Why It's Important

Glacial erosion and deposition create many landforms on Earth.

Review Vocabulary

plasticlike: not completely solid or liquid; capable of being molded or changing form

New Vocabulary

- glacier
- plucking
- till
- moraine
- outwash

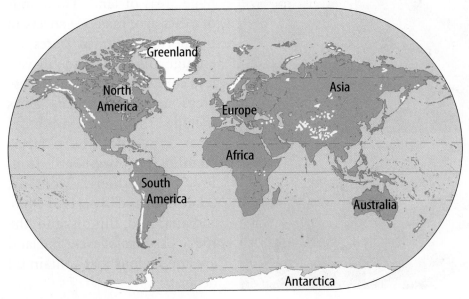

Figure 8 The white regions on this map show areas that are glaciated today.

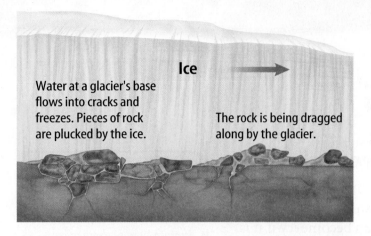

Water at a glacier's base flows into cracks and freezes. Pieces of rock are plucked by the ice.

Ice

The rock is being dragged along by the glacier.

Plucking Glaciers weather and erode solid rock. When glacial ice melts, water flows into cracks in rocks. Later, the water refreezes in these cracks, expands, and fractures the rock. Pieces of rock then are lifted out by the ice, as shown in **Figure 9.** This process, called **plucking,** results in boulders, gravel, and sand being added to the bottom and sides of a glacier.

Reading Check *What is plucking?*

Figure 9 Plucking is a process that occurs when a moving glacier picks up loosened rock particles.

Transporting and Scouring As it moves forward over land, a glacier can transport huge volumes of sediment and rock. Plucked rock fragments and sand at its base scour and scrape the soil and bedrock like sandpaper against wood, eroding the ground below even more. When bedrock is gouged deeply by rock fragments being dragged along, marks such as those in **Figure 10** are left behind. These marks, called grooves, are deep, long, parallel scars on rocks. Shallower marks are called striations (stri AY shunz). Grooves and striations indicate the direction in which the glacier moved.

Ice Depositing Sediment

Figure 10 When glaciers melt, striations or grooves can be found on the rocks beneath. These glacial grooves on Kelley's Island, Ohio, give evidence of past glacial erosion and movement.

When glaciers begin to melt, they are unable to carry much sediment. The sediment drops, or is deposited, on the land. When a glacier melts and begins to shrink back, it is said to retreat. As it retreats, a jumble of boulders, sand, clay, and silt is left behind. This mixture of different-sized sediments is called **till.** Till deposits can cover huge areas of land. Thousands of years ago, huge ice sheets in the northern United States left enough till behind to fill valleys completely and make these areas appear flat. Till areas include the wide swath of what are now wheat farms running northwestward from Iowa to northern Montana. Some farmland in parts of Ohio, Indiana, and Illinois and the rocky pastures of New England are also regions that contain till deposits.

Moraine Deposits Till also is deposited at the end of a glacier when it is not moving forward. Unlike the till that is left behind as a sheet of sediment over the land, this type of deposit doesn't cover such a wide area. Rocks and soil are moved to the end of the glacier, much like items on a grocery store conveyor belt. Because of this, a big ridge of material piles up that looks as though it has been pushed along by a bulldozer. Such a ridge is called a **moraine.** Moraines also are deposited along the sides of a glacier, as shown in **Figure 11.**

Outwash Deposits When glacial ice starts to melt, the meltwater can deposit sediment that is different from till. Material deposited by the meltwater from a glacier, most often beyond the end of the glacier, is called **outwash.** Meltwater carries sediments and deposits them in layers. Heavier sediments drop first, so bigger pieces of rock are deposited closer to the glacier. The outwash from a glacier also can form into a fan-shaped deposit when the stream of meltwater deposits sand and gravel in front of the glacier.

Reading Check *What is outwash?*

Eskers Another type of outwash deposit looks like a long, winding ridge. This deposit forms in a melting glacier when meltwater forms a river within the ice, as shown in the diagram in **Figure 12.** This river carries sand and gravel and deposits them within its channel. When the glacier melts, a winding ridge of sand and gravel, called an esker (ES kur), is left behind. An esker is shown in the photograph in **Figure 12.**

Figure 11 Moraines are forming along the sides of this glacier. Unlike the moraines that form at the ends of glaciers, these moraines form as rock and sediment fall from nearby slopes.

Figure 12 Eskers are glacial deposits formed by meltwater.

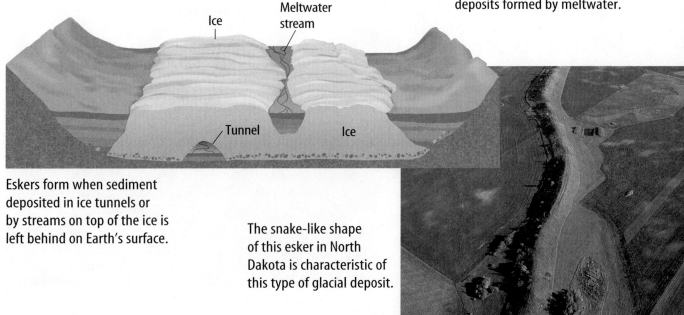

Ice

Meltwater stream

Tunnel

Ice

Eskers form when sediment deposited in ice tunnels or by streams on top of the ice is left behind on Earth's surface.

The snake-like shape of this esker in North Dakota is characteristic of this type of glacial deposit.

Continental Glaciers

The two types of glaciers are continental glaciers and valley glaciers. Today, continental glaciers, like the one in **Figure 13** cover ten percent of Earth, mostly near the poles in Antarctica and Greenland. These continental glaciers are huge masses of ice and snow. Continental glaciers are thicker than some mountain ranges. Glaciers make it impossible to see most of the land features in Antarctica and Greenland.

 **Reading Check** *In what regions on Earth would you expect to find continental glaciers?*

Figure 13 Continental glaciers and valley glaciers are agents of erosion and deposition. This continental glacier covers a large area in Antarctica.

Climate Changes In the past, continental glaciers covered as much as 28 percent of Earth. **Figure 14** shows how much of North America was covered by glaciers during the most recent ice advance. These periods of widespread glaciation are known as ice ages. Extensive glaciers have covered large portions of Earth many times over the last 2 million to 3 million years. During this time, glaciers advanced and retreated many times over much of North America. The average air temperature on Earth was about 5°C lower during these ice ages than it is today. The last major advance of ice reached its maximum extent about 18,000 years ago. After this last advance of glaciers, the ends of the ice sheets began to recede, or move back, by melting.

Figure 14 This map shows how much of North America was covered by continental glaciers about 18,000 years ago.
Observe *Was your location covered? If so, what evidence of glaciers does your area show?*

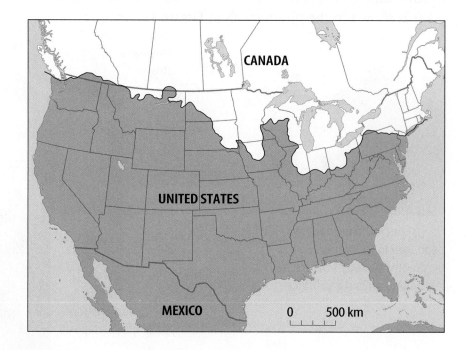

Valley Glaciers

Valley glaciers occur even in today's warmer global climate. In the high mountains where the average temperature is low enough to prevent snow from melting during the summer, valley glaciers grow and creep along. **Figure 15** shows valley glaciers in Africa.

Evidence of Valley Glaciers If you visit the mountains, you can tell whether valley glaciers ever existed there. You might look for striations, then search for evidence of plucking. Glacial plucking often occurs near the top of a mountain where a glacier is mainly in contact with solid rock. Valley glaciers erode bowl-shaped basins, called cirques (SURKS), into the sides of mountains. If two valley glaciers side by side erode a mountain, a long ridge called an arête (ah RAYT) forms between them. If valley glaciers erode a mountain from several directions, a sharpened peak called a horn might form. **Figure 16** shows some features formed by valley glaciers.

Valley glaciers flow down mountain slopes and along valleys, eroding as they go. Valleys that have been eroded by glaciers have a different shape from those eroded by streams. Stream-eroded valleys are normally V-shaped. Glacially eroded valleys are U-shaped because a glacier plucks and scrapes soil and rock from the sides as well as from the bottom. A large U-shaped valley and smaller hanging valleys are illustrated in **Figure 16.**

Figure 15 Valley glaciers, like these on Mount Kilimanjaro in north Tanzania, Africa, form between mountain peaks that lie above the snow line, where snow lasts all year.

Figure 16 Valley glaciers transform the mountains over which they pass.

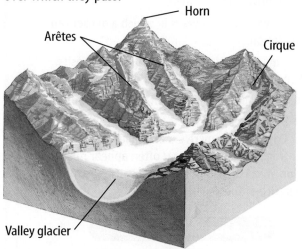

Bowl-shaped basins called cirques form by erosion at the start of a valley glacier. Arêtes form where two adjacent valley glaciers meet and erode a long, sharp ridge. Horns are sharpened peaks formed by glacial action in three or more cirques.

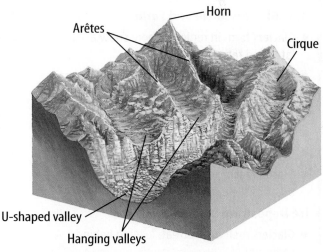

U-shaped valleys result when valley glaciers move through regions once occupied by streams. A tributary glacial valley whose mouth is high above the floor of the main valley is called a hanging valley. The discordance between the different valley floors is due to the greater erosive power of the trunk glacier.

Figure 17 Sand and gravel deposits left by glaciers are important starting materials for the construction of roadways and buildings.

Topic: Glacial Deposits

Visit bookg.msscience.com for Web links to information about uses of glacial deposits.

Activity List various uses of glacial deposits and name the methods of removing this material.

Importance of Glaciers

Glaciers have had a profound effect on Earth's surface. They have eroded mountaintops and transformed valleys. Vast areas of the continents have sediments that were deposited by great ice sheets. Today, glaciers in polar regions and in mountains continue to change the surface features of Earth.

In addition to changing the appearance of Earth's surface, glaciers leave behind sediments that are economically important, as illustrated in **Figure 17.** The sand and gravel deposits from glacial outwash and eskers are important resources. These deposits are excellent starting materials for the construction of roads and buildings.

section 2 review

Summary

How Glaciers Move and Form

- Glaciers form in regions where snow accumulates and remains year round.
- The weight of snow compresses the lower layers into ice and causes the ice to become plasticlike.

Ice Eroding Rock

- Glaciers are agents of erosion.
- Glacial erosion and deposition change large areas of the Earth's surface.

Ice Depositing Sediment

- Glaciers melt and retreat, leaving behind sediment.
- Forms of glacial deposits include till, moraines, outwash deposits and eskers.

Continental and Valley Glaciers

- Continental and valley glaciers are the two types of glaciers.

Self Check

1. **Describe** how glaciers move.
2. **Identify** two common ways in which a glacier can cause erosion.
3. **Determine** Till and outwash are glacial deposits. Explain how till and outwash are different.
4. **Discuss** How do moraines form? What are moraines made of?
5. **Think Critically** Many rivers and lakes that receive water from glacial meltwater often appear milky blue in color. What do you think might cause the milk appearance of these waters?

Applying Skills

6. **Recognize Cause and Effect** Since 1900, the Alps have lost 50 percent of their ice caps, and New Zealand's glaciers have shrunk by 26 percent. Describe what you think some causes and effects of this glacial melting have been.

 bookg.msscience.com/self_check_quiz

GLACIAL GROOVING

Throughout the world's mountainous regions, 200,000 valley glaciers are moving in response to gravity.

◉ Real-World Question

How is the land affected when a valley glacier moves downslope?

Goals
■ **Compare** stream and glacial valleys.

Materials
sand   *wood block
large plastic or   metric ruler
    metal tray   overhead light source
*stream table   with reflector
ice block   *Alternate materials
books (2 or 3)

Safety Precautions

WARNINGS: *Do not pour sand down the drain. Make sure source is plugged into a GFI electrical outlet. Do not touch light source—it may be hot.*

◉ Procedure

1. Set up the large tray of sand as shown above. Place books under one end of the tray to make a slope.

2. Cut a narrow riverlike channel through the sand. Measure and record its width and depth in a table similar to the one shown. Draw a sketch that includes these measurements.

3. Position the overhead light source to shine on the channel as shown.

4. Force the ice block into the channel at the upper end of the tray.

5. Gently push the ice along the channel until it's halfway between the top and bottom of the tray, and directly under the light.

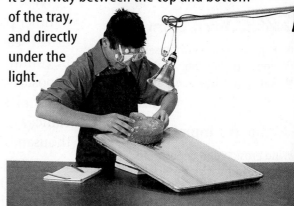

6. Turn on the light and allow the ice to melt. Record what happens.

7. Record the width and depth of the ice channel in the table. Make a scale drawing.

◉ Conclude and Apply

1. **Explain** how you can determine the direction that a glacier traveled from the location of deposits.

2. **Explain** how you can determine the direction of glacial movement from sediments deposited by meltwater.

3. **Describe** how valley glaciers affect the surface over which they move.

Glacier Data			
Sample Data	Width (cm)	Depth (cm)	Observations
Original channel	1–2	3	Stream channel looked V-shaped
Glacier channel		*Do not write in this book.*	
Meltwater channel			

Wind

as you read

What **You'll Learn**

- **Explain** how wind causes deflation and abrasion.
- **Recognize** how loess and dunes form.

Why **It's Important**

Wind erosion and deposition change landscapes, especially in dry climates.

Review Vocabulary

friction: force that opposes the motion of an object when the object is in contact with another object or surface

New Vocabulary

- deflation
- abrasion
- loess
- dune

Figure 18 The odd shape of this boulder was produced by wind abrasion.

Wind Erosion

When air moves, it picks up loose material and transports it to other places. Air differs from other erosional forces because it usually cannot pick up heavy sediments. Unlike rivers that move in confined places like channels and valleys, wind carries and deposits sediments over large areas. For example, wind is capable of picking up and carrying dust particles from fields or volcanic ash high into the atmosphere and depositing them thousands of kilometers away.

Deflation Wind erodes Earth's surface by deflation (dih FLAY shun) and abrasion (uh BRAY zhun). When wind erodes by **deflation,** it blows across loose sediment, removing small particles such as silt and sand. The heavier, coarser material is left behind.

Abrasion When windblown sediment strikes rock, the surface of the rock gets scraped and worn away by a process called **abrasion.** Abrasion, shown in **Figure 18,** is similar to sandblasting. Workers use machines that spray a mixture of sand and water under high pressure against a building. The friction wears away dirt from stone, concrete, or brick walls. It also polishes the walls of buildings by breaking away small pieces and leaving an even, smooth finish. Wind acts like a sandblasting machine, bouncing and blowing sand grains along. These sand grains strike against rocks and break off small fragments. The rocks become pitted and are worn down gradually.

Reading Check *How is wind abrasion similar to sandblasting?*

Deflation and abrasion happen to all land surfaces but occur mostly in deserts, beaches, and plowed fields. These areas have fewer plants to hold the sediments in place. When winds blow over them, they can be eroded rapidly. Grassland or pasture land have many plants that hold the soil in place, therefore there is little soil erosion caused by the wind.

Sandstorms Even when the wind blows strongly, it seldom carries sand grains higher than 0.5 m from the ground. However, sandstorms do occur. When the wind blows forcefully in the sandy parts of deserts, sand grains bounce along and hit other sand grains, causing more and more grains to rise into the air. These windblown sand grains form a low cloud just above the ground. Most sandstorms occur in deserts, but they can occur in other arid regions.

Dust Storms When soil is moist, it stays packed on the ground, but when it dries out, it can be eroded by wind. Soil is composed largely of silt- and clay-sized particles. Because these small particles weigh less than sand-sized particles of the same material, wind can move them high into the air.

Silt and clay particles are small and stick together. A faster wind is needed to lift these fine particles of soil than is needed to lift grains of sand. However, after they are airborne, the wind can carry them long distances. Where the land is dry, dust storms can cover hundreds of kilometers. These storms blow topsoil from open fields, overgrazed areas, and places where vegetation has disappeared. In the 1930s, silt and dust that was picked up in Kansas fell in New England and in the North Atlantic Ocean. Dust blown from the Sahara has been traced as far away as the West Indies—a distance of at least 6,000 km.

INTEGRATE History

Dust Bowl Poor agricultural practices and a long period of sustained drought caused the Dust Bowl of the 1930s. Research how this affected the livelihood of the people of the southern plains.

Applying Science

What factors affect wind erosion?

Many factors compound the effects of wind erosion. But can anything be done to minimize erosion?

Identifying the Problem

Wind velocity and duration, the size of sediment particles, the size of the area subjected to the wind, and the amount of vegetation present all affect how much soil is eroded by wind. The table shows different combinations of these factors. It also includes an erosion rating that depends upon what factors pertain to an area.

Factors That Affect Wind Erosion

Factor	Descriptions				
Wind velocity	high	high	low	low	low
Duration of wind	long	long	short	long	long
Particle size	coarse	medium	coarse	coarse	medium
Surface area	large	large	small	small	large
Amount of vegetation	high	low	high	high	high
Erosion rating	some	a lot	a little	some	?

Solving the Problem
1. Looking at the table, can you figure out which factors increase and which factors decrease the amount of erosion?
2. From what you've discovered, can you estimate the missing erosion rating?

Figure 19 Rows of grasses and rocks were installed on these dunes in Qinghai, China, to reduce wind erosion.

Reducing Wind Erosion

 As you've learned, wind erosion is most common where there are no plants to protect the soil. Therefore, one of the best ways to slow or stop wind erosion is to plant vegetation. This practice helps conserve soil and protect valuable farmland.

Windbreaks People in many countries plant vegetation to reduce wind erosion. For centuries, farmers have planted trees along their fields to act as windbreaks that prevent soil erosion. As the wind hits the trees, its energy of motion is reduced. It no longer is able to lift particles.

In one study, a thin belt of cottonwood trees reduced the effect of a 25-km/h wind to about 66 percent of its normal speed, or to about 16.5 km/h. Tree belts also trap snow and hold it on land. This increases the moisture level of the soil, which helps prevent further erosion.

Roots Along many seacoasts and deserts, vegetation is planted to reduce erosion. Plants with fibrous root systems, such as grasses, work best at stopping wind erosion. Grass roots are shallow and slender with many fibers. They twist and turn between particles in the soil and hold it in place.

Planting vegetation is a good way to reduce the effects of deflation and abrasion. Even so, if the wind is strong and the soil is dry, nothing can stop erosion completely. **Figure 19** shows a project designed to decrease wind erosion.

Deposition by Wind

Sediments blown away by wind eventually are deposited. Over time, these windblown deposits develop into landforms, such as dunes and accumulations of loess.

Loess Some examples of large deposits of windblown sediments are found near the Mississippi and Missouri Rivers. These wind deposits of fine-grained sediments known as **loess** (LES) are shown in **Figure 20.** Strong winds that blew across glacial outwash areas carried the sediments and deposited them. The sediments settled on hilltops and in valleys. Once there, the particles packed together, creating a thick, unlayered, yellowish-brown-colored deposit. Loess is as fine as talcum powder. Many farmlands of the midwestern United States have fertile soils that developed from loess deposits.

Dunes Do you notice what happens when wind blows sediments against an obstacle such as a rock or a clump of vegetation? The wind sweeps around or over the obstacle. Like a river, air drops sediment when its energy decreases. Sediment starts to build up behind the obstacle. The sediment itself then becomes an obstacle, trapping even more material. If the wind blows long enough, the mound will become a dune, as shown in **Figure 21.** A **dune** (DOON) is a mound of sediments drifted by the wind.

Reading Check *What is a dune?*

Dunes are common in desert regions. You also can see sand dunes along the shores of oceans, seas, or lakes. If dry sediments exist in an area where prevailing winds or sea breezes blow daily, dunes build up. Sand or other sediment will continue to build up and form a dune until the sand runs out or the obstruction is removed. Some desert sand dunes can grow to 100 m high, but most are much shorter.

Moving Dunes A sand dune has two sides. The side facing the wind has a gentler slope. The side away from the wind is steeper. Examining the shape of a dune tells you the direction from which the wind usually blows.

Unless sand dunes are planted with grasses, most dunes move, or migrate away from the direction of the wind. This process is shown in **Figure 22.** Some dunes are known as traveling dunes because they move rapidly across desert areas. As they lose sand on one side, they build it up on the other.

Figure 20 This sediment deposit is composed partially of windblown loess.

Figure 21 Loose sediment of any type can form a dune if enough of it is present and an obstacle lies in the path of the wind.

Figure 22

Sand blown loose from dry desert soil often builds up into dunes. A dune may begin to form when windblown sand is deposited in the sheltered area behind an obstacle, such as a rock outcrop. The sand pile grows as more grains accumulate. As shown in the diagram at right, dunes are mobile, gradually moved along by wind.

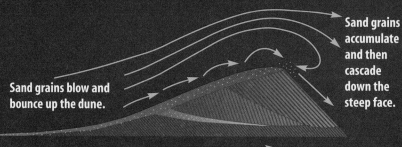

Sand grains blow and bounce up the dune.

Sand grains accumulate and then cascade down the steep face.

Dune migration

▲ A dune migrates as sand blows up its sloping side and then cascades down the steeper side. Gradually, a dune moves forward—in the same direction that the wind is blowing—as sand, lost from one side, piles up on the other side.

▲ Dunes are made of sediments eroded from local materials. Although many dunes are composed of quartz and feldspar, the brilliant white dunes in White Sands National Park, New Mexico, are made of gypsum.

▲ Deserts may expand when humans move into the transition zone between habitable land and desert. Here, villagers in Mauritania in northwestern Africa shovel the sand that encroaches on their schoolhouse daily.

◀ The dunes at left are coastal dunes from the Laguna Madre region of South Texas on the Gulf of Mexico. Note the vegetation in the photo, which has served as an obstacle to trap sand.

Dune Shape The shape of a dune depends on the amount of sand or other sediment available, the wind speed and direction, and the amount of vegetation present. One common dune shape is a crescent-shaped dune known as a barchan (BAR kun) dune. The open side of a barchan dune faces the direction that the wind is blowing. When viewed from above, the points of the crescent are directed downwind. This type of dune forms on hard surfaces where the sand supply is limited.

Another common type of dune, called a transverse dune, forms where sand is abundant. Transverse dunes are so named because the long directions of these dunes are perpendicular to the general wind direction. In regions where the wind direction changes, star dunes, shown in **Figure 23,** form pointed structures. Other dune forms also exist, some of which show a combination of features.

Shifting Sediments When dunes and loess form, the landscape changes. Wind, like gravity, running water, and glaciers, shapes the land. New landforms created by these agents of erosion are themselves being eroded. Erosion and deposition are part of a cycle of change that constantly shapes and reshapes the land.

Figure 23 Star dunes form in areas where the wind blows from several different directions.

section 3 review

Summary

Wind Erosion

- Air movement picks up loose material and transports it to other places.
- Deflation and abrasion happen mainly in deserts, beaches, and plowed fields.

Reducing Wind Erosion

- Planting vegetation can reduce wind erosion.
- Farmers use windbreaks to protect their crop fields from wind erosion.

Deposition by Wind

- Windblown deposits develop into landforms, such as dunes and accumulation of loess.
- Many farmlands of the midwestern United States have fertile soils that developed from loess deposits.

Self Check

1. **Compare and contrast** abrasion and deflation. Describe how they affect the surface of Earth.

2. **Explain** the differences between dust storms and sandstorms. Describe how energy of motion affects the deposition of sand and dust by these storms.

3. **Think Critically** You notice that sand is piling up behind a fence outside your apartment building. Explain why this occurs.

Applying Math

4. **Solve One-Step Equations** Between 1972 and 1992, the Sahara increased by nearly 700 km² in Mali and the Sudan. Calculate the average number of square kilometers the desert increased each year between 1972 and 1992.

LAB

Design Your Own

Blowing in the Wind

▶ Real-World Question

Have you ever played a sport outside and suddenly had the wind blow dust into your eyes? What did you do? Turn your back? Cover your eyes? How does wind pick up sediment? Why does wind pick up some sediments and leave others on the ground? What factors affect wind erosion?

▶ Form a Hypothesis

How does moisture in sediment affect the ability of wind to erode sediments? Does the speed of the wind limit the size of sediments it can transport? Form a hypothesis about how sediment moisture affects wind erosion. Form another hypothesis about how wind speed affects the size of the sediment the wind can transport.

▶ Test Your Hypothesis

Make a Plan

1. As a group, agree upon and write your hypothesis statements.

2. **List** the steps needed to test your first hypothesis. Plan specific steps and vary only one factor at a time. Then, list the steps needed to test your second hypothesis. Test only one factor at a time.

Goals

- **Observe** the effects of soil moisture and wind speed on wind erosion.
- **Design** and carry out experiments that test the effects of soil moisture and wind speed on wind erosion.

Possible Materials

flat pans (4)
fine sand (400 mL)
gravel (400 mL)
hair dryer
sprinkling can
water
28-cm × 35-cm cardboard sheets (4)
tape
mixing bowl
metric ruler
wind speed indicator

Safety Precautions

Wear your safety goggles at all times when using the hair dryer on sediments. Make sure the dryer is plugged into a GFI electrical outlet.

3. Mix the sediments in the pans. Plan how you will fold cardboard sheets and attach them to the pans to keep sediments contained.

4. **Design** data tables in your Science Journal. Use them as your group collects data.

5. **Identify** all constants, variables, and controls of the experiment. One example of a control is a pan of sediment not subjected to any wind.

Follow Your Plan

1. Make sure your teacher approves your plan before you start.

2. Carry out the experiments as planned.

3. While doing the experiments, write any observations that you or other members of your group make. Summarize your data in the data tables you designed in your Science Journal.

Sediment Movement		
Sediment	Wind Speed	Sediment Moved
Fine sand (dry)	low	
	high	
Fine sand (wet)	low	Do not write in this book.
	high	
Gravel (dry)	low	
	high	
Gravel (wet)	low	
	high	
Fine sand and gravel (dry)	low	
	high	
Fine sand and gravel (wet)	low	
	high	

▶ Analyze Your Data

1. **Compare** your results with those of other groups. Explain what might have caused any differences among the groups.

2. **Explain** the relationship that exists between the speed of the wind and the size of the sediments it transports.

▶ Conclude and Apply

1. How does energy of motion of the wind influence sediment transport? What is the general relationship between wind speed and erosion?

2. **Explain** the relationship between the sediment moisture and the amount of sediment moved by the wind.

Communicating Your Data

Design a table that summarizes the results of your experiment, and use it to explain your interpretations to others in the class.

SCIENCE Stats

Losing Against Erosion

Did you know...

...Glaciers, one of nature's most powerful erosional forces, can move more than 30 m per day. In one week, a fast-moving glacier can travel the length of almost two football fields. Glaciers such as these are unusual—most move less than 10 cm per day.

...Some sand dunes migrate as much as 30 m per year. In a coastal region of France, traveling dunes have buried farms and villages. The dunes were halted by anti-erosion practices, such as planting grass in the sand and growing a barrier of trees between the dunes and farmland.

Applying Math If a sand dune is traveling at 30 m per year, how many meters does it travel in one month?

...In 1959, an earthquake triggered a mass movement in Madison River Canyon, Montana. About 21 million km^3 of rock and soil slid down the canyon at an estimated 160 km/h. This type of mass movement of earth is called a rock slide.

Find Out About It

Visit bookg.msscience.com/science_stats **to learn about landslides. When is a landslide called a mudflow? In which U.S. states are mudflows most likely to occur?**

Reviewing Main Ideas

Section 1 Erosion by Gravity

1. Erosion is the process that picks up and transports sediment.

2. Deposition occurs when an agent of erosion loses its energy and can no longer carry its load of sediment.

3. Slump, creep, rock slides, and mudflows are all mass movements caused by gravity.

Section 2 Glaciers

1. Glaciers are powerful agents of erosion. As water freezes and thaws in cracks, it breaks off pieces of surrounding rock. These pieces then are incorporated into glacial ice by plucking.

2. As sediment embedded in the base of a glacier moves across the land, grooves and striations form. Glaciers deposit two kinds of material—till and outwash.

Section 3 Wind

1. Deflation occurs when wind erodes only fine-grained sediments, leaving coarse sediments behind.

2. The pitting and polishing of rocks and grains of sediment by windblown sediment is called abrasion.

3. Wind deposits include loess and dunes. Loess consists of fine-grained particles such as silt and clay. Dunes form when windblown sediments accumulate behind an obstacle.

Visualizing Main Ideas

Copy and complete the following concept map on erosional forces. Use the following terms and phrases: striations, leaning trees and structures, curved scar on slope, deflation, *and* mudflows.

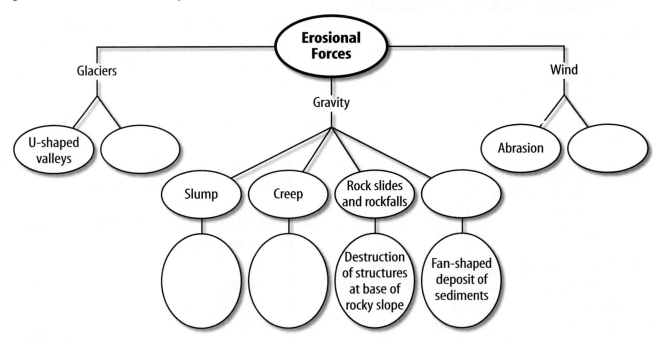

Using Vocabulary

abrasion p. 76	loess p. 79
creep p. 66	mass movement p. 65
deflation p. 76	moraine p. 71
deposition p. 65	outwash p. 71
dune p. 79	plucking p. 70
erosion p. 64	slump p. 65
glacier p. 69	till p. 70

Each phrase below describes a vocabulary word from the list. In your Science Journal, write the term that matches each description.

1. loess, dunes, and moraines are examples

2. slowest mass movement

3. ice picking up pieces of rock

4. much like sandblasting

5. gravity transport of material downslope

6. sand and gravel deposited by meltwater

7. glacial deposit composed of sediment with many sizes and shapes

Checking Concepts

Choose the word or phrase that best answers the question.

8. Which term is an example of a feature created by deposition?
   A) cirque   C) striation
   B) abrasion   D) dune

9. The best plants for reducing wind erosion have what type of root system?
   A) taproot   C) fibrous
   B) striated   D) sheet

10. What does a valley glacier create at the point where it starts?
    A) esker   C) till
    B) moraine   D) cirque

Use the photo below to answer question 11.

11. Which of the following is suggested by leaning trees curving upright on a hillside?
    A) abrasion   C) slump
    B) creep   D) mudflow

12. What shape do glacier-created valleys have?
    A) V-shape   C) U-shape
    B) L-shape   D) S-shape

13. Which is formed by glacial erosion?
    A) eskers   C) moraines
    B) arêtes   D) warmer climate

14. What type of wind erosion leaves pebbles and boulders behind?
    A) deflation   C) abrasion
    B) loess   D) sandblasting

15. What is a ridge formed by deposition of till called?
    A) striation   C) cirque
    B) esker   D) moraine

16. What is the material called that is deposited by meltwater beyond the end of a glacier?
    A) esker
    B) cirque
    C) outwash
    D) moraine

Thinking Critically

17. Explain how striations can give information about the direction that a glacier moved.

18. Describe how effective a retaining wall made of fine wire mesh would be against erosion.

19. Determine what can be done to prevent the migration of beach dunes.

20. Recognize Cause and Effect A researcher finds evidence of movement of ice within a glacier. Explain how this movement could occur.

21. Think Critically The end of a valley glacier is at a lower elevation than its point of origin is. How does this help explain melting at its end while snow and ice still are accumulating where it originated?

22. Make and Use Tables Make a table to contrast continental and valley glaciers.

23. Concept Map Copy and complete the events-chain concept map below to show how a sand dune forms. Use the terms and phrases: *sand accumulates, dune, dry sand,* and *obstruction traps.*

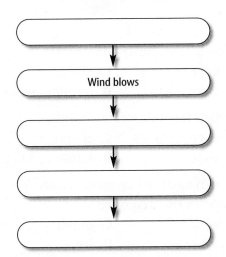

Wind blows

24. Form a Hypothesis Hypothesize why silt in loess deposits is transported farther than sand in dune deposits.

25. Test a Hypothesis Explain how to test the effect of glacial thickness on a glacier's ability to erode.

26. Classify the following as erosional or depositional features: loess, cirque, U-shaped valley, sand dune, abraded rock, striation, and moraine.

Performance Activities

27. Poster Make a poster with magazine photos showing glacial features in North America. Add a map to locate each feature.

28. Design an experiment to see how the amount of moisture added to sediments affects mass movement. Keep all variables constant except the amount of moisture in the sediment. Try your experiment.

Applying Math

29. Slope Gravity is a very powerful erosional force. This means the steeper a slope is, the more soil will move. A person can calculate how steep a slope is by using the height (rise) divided by the length (run). This answer is then multiplied by 100 to get percent slope. If you had a slope 15 m high and 50 m long, what would be the percent slope?

Slope

15 m (rise)

50 m (run)

30. Traveling Sand A sand dune can travel up to 30 m per year. How far does the sand dune move per day?

Part 1 | Multiple Choice

Record your answers on the answer sheet provided by your teacher or on a sheet of paper.

Use the illustration below to answer question 1.

1. Which type of mass movement is shown above?
   - **A.** slump
   - **B.** creep
   - **C.** rock slide
   - **D.** mudflow

2. Which term refers to sediment that is deposited by glacier ice?
   - **A.** outwash
   - **B.** till
   - **C.** loess
   - **D.** esker

3. During which process does wind pick up fine sediment?
   - **A.** abrasion
   - **B.** mass movement
   - **C.** deflation
   - **D.** deposition

4. On which of the following continents do continental glaciers exist today?
   - **A.** Antarctica
   - **B.** Africa
   - **C.** Australia
   - **D.** Europe

Test-Taking Tip

Take Your Time Stay focused during the test and don't rush, even if you notice that other students are finishing the test early.

5. Which forms when a rock in glacier ice slides over Earth's surface?
   - **A.** moraine
   - **B.** esker
   - **C.** horn
   - **D.** groove

6. What causes sediment and rock to move to lower elevations through time?
   - **A.** sunlight
   - **B.** plant roots
   - **C.** gravity
   - **D.** dust storms

7. Which can reduce wind erosion?
   - **A.** windbreaks
   - **B.** dunes
   - **C.** eskers
   - **D.** horns

8. Which consists of fine-grained, wind-blown sediment?
   - **A.** moraine
   - **B.** loess
   - **C.** till
   - **D.** rock fall

Use the diagram below to answer questions 9–11.

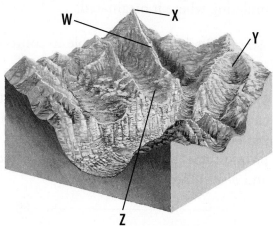

9. Which term describes point X?
   - **A.** horn
   - **B.** arête
   - **C.** cirque
   - **D.** hanging valley

10. Which term describes point Z?
    - **A.** horn
    - **B.** arête
    - **C.** cirque
    - **D.** hanging valley

11. Which agent of erosion created the landscape in the diagram?
    - **A.** wind
    - **B.** water
    - **C.** gravity
    - **D.** ice

Part 2 | Short Response/Grid In

Record your answers on the answer sheet provided by your teacher or on a separate sheet of paper.

12. Give three examples of erosion? How does it affect Earth's surface?

13. What is deposition? Give three examples of how it changes Earth's surface.

14. How is a rock fall different from a rock slide? Use a labeled diagram to support your answer.

15. Explain how a glacier can erode the land, and then describe three forms of glacial deposition.

The graph below shows data about how much water flows through a stream. The stream is fed by glacial meltwater. Use the graph to answer questions 16–18.

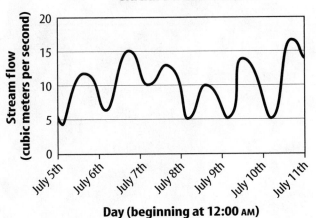

Glacial Stream Flow

16. What were the lowest and highest amounts of stream flow on July 6th?

17. What were the lowest and highest amounts of stream flow on July 8th?

18. Notice that these data were obtained in July. Explain why the amount of stream flow from a glacier would vary each day. Give three examples to support your reasoning.

Part 3 | Open Ended

Record your answers on a sheet of paper.

Use the diagram below to answer questions 19–22.

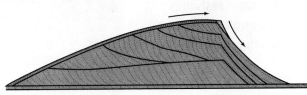

19. Describe the process that causes sand to move up the less steep side of the dune. Use a labeled diagram to support your answer.

20. Why does sand move down the steeper side of the dune? Use a labeled diagram to support your answer.

21. Design three time-lapse illustrations to show how sand dunes move across land.

22. Describe three ways to slow down the movement of sand dunes.

23. What types of damage are caused by landslides? How are people affected by landslides physically and economically?

24. Give three ways that damage from landslides can be reduced.

25. What is a dust storm? Where would you expect dust storms to occur? Give two safety suggestions for people caught in a dust storm.

26. Describe two ways that glacier ice can move across the surface.

27. How is creep different from most other types of mass movement? Explain the forces that cause creep, as well as the effect of creep.

28. Create a chart to show how continental glaciers are different from valley glaciers. Include their causes, physical features, and geological effects on the land.

chapter
4

The BIG Idea

Surface water reshapes Earth through erosion and deposition processes.

SECTION 1
Surface Water
Main Idea As water moves across Earth's surface, it erodes soil and rock from one location and deposits the sediment in another.

SECTION 2
Groundwater
Main Idea Water that soaks into the ground becomes part of a system that can include unique forms of erosion and deposition.

SECTION 3
Ocean Shoreline
Main Idea Waves, currents, and tides reshape shorelines through erosion and deposition processes.

Water Erosion and Deposition

Nature's Sculptor

Bryce Canyon National Park in Utah is home to the Hoodoos— tall, column-like formations. They were made by one of the most powerful forces on Earth—moving water. In this chapter you will learn how moving water shapes Earth's surface.

Science Journal What might have formed the narrowing of each Hoodoo? What will happen if this narrowing continues?

Start-Up Activities

Model How Erosion Works

Moving water has great energy. Sometimes rainwater falls softly and soaks slowly into soil. Other times it rushes down a slope with tremendous force and carries away valuable topsoil. What determines whether rain soaks into the ground or runs off and wears away the surface?

1. Place an aluminum pie pan on your desktop.
2. Put a pile of dry soil about 7 cm high into the pan.
3. Slowly drip water from a dropper onto the pile and observe what happens next.
4. Drip the water faster and continue to observe what happens.
5. Repeat steps 1 through 4, but this time change the slope of the hill by increasing the central pile. Start again with dry soil.
6. **Think Critically** In your Science Journal, write about the effect the water had on the different slopes.

Characteristics of Surface Water, Groundwater, and Shoreline Water Make the following Foldable to help you identify the main concepts relating to surface water, groundwater, and shoreline water.

STEP 1 Fold the top of a vertical piece of paper down and the bottom up to divide the paper into thirds.

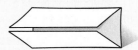

STEP 2 Turn the paper horizontally; unfold and label the three columns as shown.

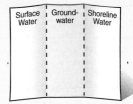

Read for Main Ideas As you read the chapter, list the concepts relating to surface water, groundwater, shoreline water.

Preview this chapter's content and activities at bookg.msscience.com

Get Ready to Read

① Learn It! Make connections between what you read and what you already know. Connections can be based on personal experiences (text-to-self), what you have read before (text-to-text), or events in other places (text-to-world).

As you read, ask connecting questions. Are you reminded of a personal experience? Have you read about the topic before? Did you think of a person, a place, or an event in another part of the world?

② Practice It! Read the excerpt below and make connections to your own knowledge and experience.

> **Have you ever seen or been in a flood? What do you think causes floods?**

> **What have you read about floods in other chapters? What types of weather events cause floods?**

Sometimes heavy rains or sudden melting of snow can cause large amounts of water to enter a river system. What happens when a river system has too much water in it? The water needs to go somewhere, and out and over the banks is the only place it can go. A river that overflows its banks can bring disaster by flooding homes or washing away bridges or crops.

—*from page 100*

> **How can people protect themselves from floods? How do people unintentionally contribute to the damage caused by floods?**

③ Apply It! As you read this chapter, choose five words or phrases that make a connection to something you already know.

Target Your Reading

Reading Tip

Make connections with memorable events, places, or people in your life. The better the connection, the more likely you will remember.

Use this to focus on the main ideas as you read the chapter.

1 **Before you read** the chapter, respond to the statements below on your worksheet or on a numbered sheet of paper.

- Write an **A** if you **agree** with the statement.
- Write a **D** if you **disagree** with the statement.

2 **After you read** the chapter, look back to this page to see if you've changed your mind about any of the statements.

- If any of your answers changed, explain why.
- Change any false statements into true statements.
- Use your revised statements as a study guide.

Sciencenline

Print out a worksheet of this page at bookg.msscience.com

Before You Read A or D		Statement	After You Read A or D
	1	Water often causes sheet erosion when it flows in thin, broad sheets.	
	2	Streams are classified as young, mature, or old.	
	3	The stages of development of a stream depend only on the actual age of the stream.	
	4	The largest drainage basin in the United States is the Grand Canyon.	
	5	Water that soaks into the ground and collects in tiny pores in underlying rock is called groundwater.	
	6	The water table below Earth's surface always remains at the same depth.	
	7	The three major forces at work on a shoreline are waves, currents, and tides.	
	8	A longshore current runs perpendicular to the shoreline.	
	9	All sand is made of the mineral quartz.	

Surface Water

as you read

What You'll Learn

- **Identify** the causes of runoff.
- **Compare** rill, gully, sheet, and stream erosion.
- **Identify** three different stages of stream development.
- **Explain** how alluvial fans and deltas form.

Why It's Important

Runoff and streams shape Earth's surface.

🔎 Review Vocabulary

erosion: transport of surface materials by agents such as gravity, wind, water, or glaciers

New Vocabulary

- • runoff
- • channel
- • sheet erosion
- • drainage basin
- • meander

Runoff

Picture this. You pour a glass of milk, and it overflows, spilling onto the table. You grab a towel to clean up the mess, but the milk is already running through a crack in the table, over the edge, and onto the floor. This is similar to what happens to rainwater when it falls to Earth. Some rainwater soaks into the ground and some evaporates, turning into a gas. The rainwater that doesn't soak into the ground or evaporate runs over the ground. Eventually, it enters streams, lakes, or the ocean. Water that doesn't soak into the ground or evaporate but instead flows across Earth's surface is called **runoff.** If you've ever spilled milk while pouring it, you've experienced something similar to runoff.

Factors Affecting Runoff What determines whether rain soaks into the ground or runs off? The amount of rain and the length of time it falls are two factors that affect runoff. Light rain falling over several hours probably will have time to soak into the ground. Heavy rain falling in less than an hour or so will run off because it cannot soak in fast enough, or it can't soak in because the ground cannot hold any more water.

Figure 1 In areas with gentle slopes and vegetation, little runoff and erosion take place. Lack of vegetation has led to severe soil erosion in some areas.

Other Factors Another factor that affects the amount of runoff is the steepness, or slope, of the land. Gravity, the attractive force between all objects, causes water to move down slopes. Water moves rapidly down steep slopes so it has little chance to soak into the ground. Water moves more slowly down gentle slopes and across flat areas. Slower movement allows water more time to soak into the ground.

Vegetation, such as grass and trees, also affects the amount of runoff. Just like milk running off the table, water will run off smooth surfaces that have little or no vegetation. Imagine a tablecloth on the table. What would happen to the milk then? Runoff slows down when it flows around plants. Slower-moving water has a greater chance to sink into the ground. By slowing down runoff, plants and their roots help prevent soil from being carried away. Large amounts of soil may be carried away in areas that lack vegetation, as shown in **Figure 1.**

Effects of Gravity When you lie on the ground and feel as if you are being held in place, you are experiencing the effects of gravity. Gravity is the attracting force all objects have for one another. The greater the mass of an object is, the greater its force of gravity is. Because Earth has a much greater mass than any of the objects on it, Earth's gravitational force pulls objects toward its center. Water runs downhill because of Earth's gravitational pull. When water begins to run down a slope, it picks up speed. As its speed increases, so does its energy. Fast-moving water, shown in **Figure 2,** carries more soil than slow-moving water does.

INTEGRATE
Career

Conservation Farmers sometimes have to farm on some kind of slope. The steeper the slope, the more erosion will occur. Not only is slope an important factor but other factors have to be considered as well. The Natural Resources Conservation Service, a government agency, studies these factors to determine soil loss from a given area. Find out what other factors this agency uses to determine soil loss.

Figure 2 During floods, the high volume of fast-moving water erodes large amounts of soil.

Figure 3 Heavy rains can remove large amounts of sediment, forming deep gullies in the side of a slope.

Water Erosion

Suppose you and several friends walk the same way to school each day through a field or an empty lot. You always walk in the same footsteps as you did the day before. After a few weeks, you've worn a path through the field. When water travels down the same slope time after time, it also wears a path. The movement of soil and rock from one place to another is called erosion.

Rill and Gully Erosion You may have noticed a groove or small ditch on the side of a slope that was left behind by running water. This is evidence of rill erosion. Rill erosion begins when a small stream forms during a heavy rain. As this stream flows along, it has enough energy to erode and carry away soil. Water moving down the same path creates a groove, called a **channel,** on the slope where the water eroded the soil. If water frequently flows in the same channel, rill erosion may change over time into another type of erosion called gully erosion.

During gully erosion, a rill channel becomes broader and deeper. **Figure 3** shows gullies that were formed when water carried away large amounts of soil.

Sheet Erosion Water often erodes without being in a channel. Rainwater that begins to run off during a rainstorm often flows as thin, broad sheets before forming rills and streams. For example, when it rains over an area, the rainwater accumulates until it eventually begins moving down a slope as a sheet, like the water flowing off the hood of the car in **Figure 4.** Water also can flow as sheets if it breaks out of its channel.

Figure 4 When water accumulates, it can flow in sheets like the water seen flowing over the hood of this car.

Floodwaters spilling out of a river can flow as sheets over the surrounding flatlands. Streams flowing out of mountains fan out and may flow as sheets away from the foot of the mountain. **Sheet erosion** occurs when water that is flowing as sheets picks up and carries away sediments.

Stream Erosion Sometimes water continues to flow along a low place it has formed. As the water in a stream moves along, it picks up sediments from the bottom and sides of its channel. By this process, a stream channel becomes deeper and wider.

The sediment that a stream carries is called its load. Water picks up and carries some of the lightweight sediments, called the suspended load. Larger, heavy particles called the bed load just roll along the bottom of the stream channel, as shown in **Figure 5.** Water can even dissolve some rocks and carry them away in solution. The different-sized sediments scrape against the bottom and sides of the channel like a piece of sandpaper. Gradually, these sediments can wear away the rock by a process called abrasion.

Figure 5 This cross section of a stream channel shows the location of the suspended load and the bed load.
Describe *how the stream carries dissolved material.*

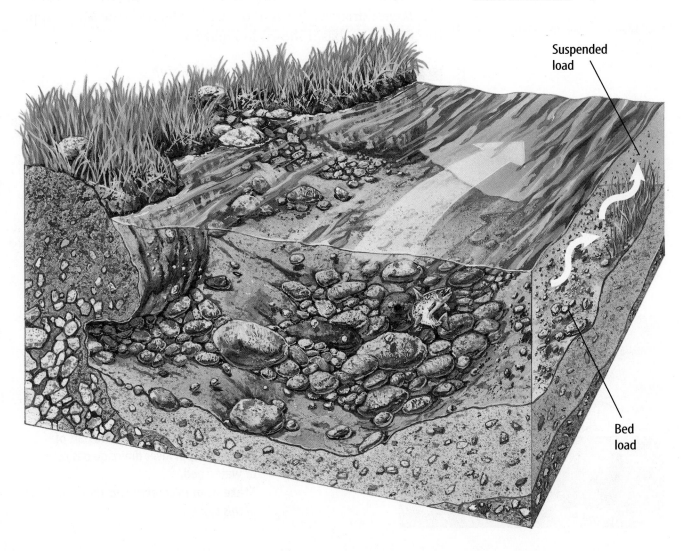

Suspended load

Bed load

River System Development

Have you spent time near a river or stream in your community? Each day, probably millions of liters of water flow through that stream. Where does all the water come from? Where is it flowing to?

River Systems Streams are parts of river systems. The water comes from rills, gullies, and smaller streams located upstream. Just as the tree in **Figure 6** is a system containing twigs, branches, and a trunk, a river system also has many parts. Runoff enters small streams, which join together to form larger streams. Larger streams come together to form rivers. Rivers grow and carry more water as more streams join.

Drainage Basins A **drainage basin** is the area of land from which a stream or river collects runoff. Compare a drainage basin to a bathtub. Water that collects in a bathtub flows toward one location—the drain. Likewise, all of the water in a river system eventually flows to one location—the main river, or trunk. The largest drainage basin in the United States is the Mississippi River drainage basin shown in **Figure 6.**

Reading Check *What is a drainage basin?*

Figure 6 River systems can be compared with the structure of a tree.

The system of twigs, branches, and trunk that make up a tree is similar to the system of streams and rivers that make up a river system.

A large number of the streams and rivers in the United States are part of the Mississippi River drainage basin, or watershed.
State *what river represents the trunk of this system.*

Stages of Stream Development

Streams come in a variety of forms. Some are narrow and swift moving, and others are wide and slow moving. Streams differ because they are in different stages of development. These stages depend on the slope of the ground over which the stream flows. Streams are classified as young, mature, or old. **Figure 8** shows how the stages come together to form a river system.

The names of the stages of development aren't always related to the actual age of a river. The New River in West Virginia is one of the oldest rivers in North America. However, it has a steep valley and flows swiftly. As a result, it is classified as a young stream.

Young Streams A stream that flows swiftly through a steep valley is a young stream. A young stream may have white-water rapids and waterfalls. Water flowing through a steep channel with a rough bottom has a high level of energy and erodes the stream bottom faster than its sides.

Mature Streams The next stage in the development of a stream is the mature stage. A mature stream flows more smoothly through its valley. Over time, most of the rocks in the streambed that cause waterfalls and rapids are eroded by running water and the sediments it carries.

Erosion is no longer concentrated on the bottom in a mature stream. A mature stream starts to erode more along its sides, and curves develop. These curves form because the speed of the water changes throughout the width of the channel.

Water in a shallow area of a stream moves slower because it drags along the bottom. In the deeper part of the channel, the water flows faster. If the deep part of the channel is next to one side of the river, water will erode that side and form a slight curve. Over time, the curve grows to become a broad arc called a **meander** (mee AN dur), as shown in **Figure 7.**

The broad, flat valley floor formed by a meandering stream is called a floodplain. When a stream floods, it often will cover part or all of the floodplain.

Figure 7 A meander is a broad bend in a river or stream. As time passes, erosion of the outer bank increases the bend.

Figure 8

Although no two streams are exactly alike, all go through three main stages—young, mature, and old—as they flow from higher to lower ground. A young stream, below, surging over steep terrain, moves rapidly. In a less steep landscape, right, a mature stream flows more smoothly. On nearly level ground, the stream—considered old—winds leisurely through its valley. The various stages of a stream's development are illustrated here.

Waterfall

Rapids

A A young stream begins at a source—here, a melting mountain glacier. From its source, the stream flows swiftly downhill, cutting a narrow valley.

B A mature stream flows smoothly through its valley. Mature streams often develop broad curves called meanders.

C Old streams flow through broad, flat floodplains. Near its mouth, the stream gradually drops its load of silt. This sediment forms a delta, an area of flat, fertile land extending into the ocean.

Oxbow lake

Science Online

Topic: Classification of Rivers

Visit bookg.msscience.com for Web links to information about major rivers in the United States. Classify two of these streams as young, mature, or old.

Activity Write a paragraph about why you think the Colorado River is a young, mature, or old stream.

Old Streams The last stage in the development of a stream is the old stage. An old stream flows smoothly through a broad, flat floodplain that it has deposited. South of St. Louis, Missouri, the lower Mississippi River is in the old stage.

Major river systems, such as the Mississippi River, usually contain streams in all stages of development. In the upstream portion of a river system, you find whitewater streams moving swiftly down mountains and hills. At the bottom of mountains and hills, you find streams that start to meander and are in the mature stage of development. These streams meet at the trunk of the drainage basin and form a major river.

Reading Check *How do old streams differ from young streams?*

Too Much Water

Sometimes heavy rains or a sudden melting of snow can cause large amounts of water to enter a river system. What happens when a river system has too much water in it? The water needs to go somewhere, and out and over the banks is the only choice. A river that overflows its banks can bring disaster by flooding homes or washing away bridges or crops.

Dams and levees are built in an attempt to prevent this type of flooding. A dam is built to control the water flow downstream. It may be built of soil, sand, or steel and concrete. Levees are mounds of earth that are built along the sides of a river. Dams and levees are built to prevent rivers from overflowing their banks. Unfortunately, they do not stop the water when flooding is great. This was the case in 1993 when heavy rains caused the Mississippi River to flood parts of nine midwestern states. Flooding resulted in billions of dollars in property damage. **Figure 9** shows some of the damage caused by this flood.

As you have seen, floods can cause great amounts of damage. But at certain times in Earth's past, great floods have completely changed the surface of Earth in a large region. Such floods are called catastrophic floods.

Figure 9 Flooding causes problems for people who live along major rivers. Floodwater broke through a levee during the Mississippi River flooding in 1993.

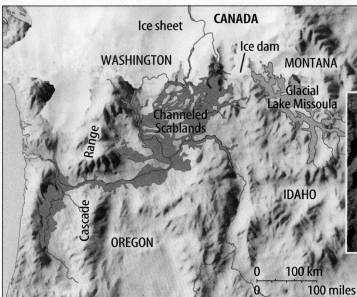

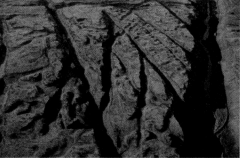

Figure 10 The Channeled Scablands formed when Lake Missoula drained catastrophically.

These channels were formed by the floodwaters.

Catastrophic Floods During Earth's long history, many catastrophic floods have dramatically changed the face of the surrounding area. One catastrophic flood formed the Channeled Scablands in eastern Washington State, shown here in **Figure 10.** A vast lake named Lake Missoula covered much of western Montana. A natural dam of ice formed this lake. As the dam melted or was eroded away, tremendous amounts of water suddenly escaped through what is now the state of Idaho into Washington. In a short period of time, the floodwater removed overlying soil and carved channels into the underlying rock, some as deep as 50 m. Flooding occurred several more times as the lake refilled with water and the dam broke loose again. Scientists say the last such flood occurred about 13,000 years ago.

Deposition by Surface Water

You know how hard it is to carry a heavy object for a long time without putting it down. As water moves throughout a river system, it loses some of its energy of motion. The water can no longer carry some of its sediment. As a result, it drops, or is deposited, to the bottom of the stream.

Some stream sediment is carried only a short distance. In fact, sediment often is deposited within the stream channel itself. Other stream sediment is carried great distances before being deposited. Sediment picked up when rill and gully erosion occur is an example of this. Water usually has a lot of energy as it moves down a steep slope. When water begins flowing on a level surface, it slows, loses energy, and deposits its sediment. Water also loses energy and deposits sediment when it empties into an ocean or lake.

Observing Runoff Collection

Procedure

1. Put a plastic **rain gauge** into a narrow **drinking glass** and place the glass in the **sink.**
2. Fill a plastic **sprinkling can** with **water.**
3. Hold the sprinkling can one-half meter above the sink for 30 s.
4. Record the amount of water in the rain gauge.
5. After emptying the rain gauge, place a **plastic funnel** into the rain gauge and sprinkle again for 30 s.
6. Record the amount of water in the gauge.

Analysis

Explain how a small amount of rain falling on a drainage basin can have a big effect on a river or stream.

Figure 11 This satellite image of the Nile River Delta in Egypt shows the typical triangular shape. The green color shows areas of vegetation.

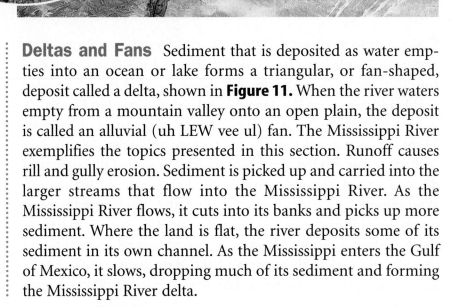

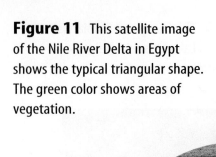

Agriculture is important on the Nile Delta.

Deltas and Fans Sediment that is deposited as water empties into an ocean or lake forms a triangular, or fan-shaped, deposit called a delta, shown in **Figure 11.** When the river waters empty from a mountain valley onto an open plain, the deposit is called an alluvial (uh LEW vee ul) fan. The Mississippi River exemplifies the topics presented in this section. Runoff causes rill and gully erosion. Sediment is picked up and carried into the larger streams that flow into the Mississippi River. As the Mississippi River flows, it cuts into its banks and picks up more sediment. Where the land is flat, the river deposits some of its sediment in its own channel. As the Mississippi enters the Gulf of Mexico, it slows, dropping much of its sediment and forming the Mississippi River delta.

section ① review

Summary

Runoff

- Rainwater that doesn't soak into the ground or evaporate becomes runoff.
- Slope of land and vegetation affect runoff.

Water Erosion

- Water flowing over the same slope causes rills and gullies to form.

River System Development

- A drainage basin is an area of land from which a stream or river collects runoff.

Self Check

1. **Explain** how the slope of an area affects runoff.
2. **Compare and contrast** rill and gully erosion.
3. **Describe** the three stages of stream development.
4. **Think Critically** How is a stream's rate of flow related to the amount of erosion it causes? How is it related to the size of the sediments it deposits?

Applying Skills

5. **Compare and contrast** the formation of deltas and alluvial fans.

Science Online bookg.msscience.com/self_check_quiz

Groundwater

Groundwater Systems

What would have happened if the spilled milk in Section 1 ran off the table onto a carpeted floor? It probably would have quickly soaked into the carpet. Water that falls on Earth can soak into the ground just like the milk into the carpet.

Water that soaks into the ground becomes part of a system, just as water that stays above ground becomes part of a river system. Soil is made up of many small rock and mineral fragments. These fragments are all touching one another, as shown in **Figure 12,** but some empty space remains between them. Holes, cracks, and crevices exist in the rock underlying the soil. Water that soaks into the ground collects in these pores and empty spaces and becomes part of what is called **groundwater.**

How much of Earth's water do you think is held in the small openings in rock? Scientists estimate that 14 percent of all freshwater on Earth exists as groundwater. This is almost 30 times more water than is contained in all of Earth's lakes and rivers.

as you read

What You'll Learn

- **Recognize** the importance of groundwater.
- **Describe** the effect that soil and rock permeability have on groundwater movement.
- **Explain** how groundwater dissolves and deposits minerals.

Why It's Important

The groundwater system is an important source of your drinking water.

Review Vocabulary
pore: a small, or minute, opening in rock or soil

New Vocabulary
- groundwater
- permeable
- impermeable
- aquifer
- water table
- spring
- geyser
- cave

Figure 12 Soil has many small, connected pores that are filled with water when soil is wet.

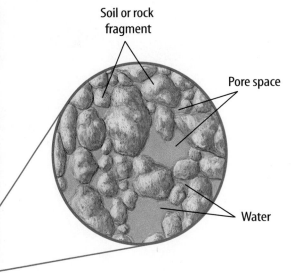

Soil or rock fragment

Pore space

Water

Permeability A groundwater system is similar to a river system. However, instead of having channels that connect different parts of the drainage basin, the groundwater system has connecting pores. Soil and rock are **permeable** (PUR mee uh bul) if the pore spaces are connected and water can pass through them. Sandstone is an example of a permeable rock.

Soil or rock that has many large, connected pores is permeable. Water can pass through it easily. However, if a rock or sediment has few pore spaces or they are not well connected, then the flow of groundwater is blocked. These materials are **impermeable,** which means that water cannot pass through them. Granite has few or no pore spaces at all. Clay has many small pore spaces, but the spaces are not well connected.

✓ **Reading Check** *How does water move through permeable rock?*

Groundwater Movement How deep into Earth's crust does groundwater go? **Figure 13** shows a model of a groundwater system. Groundwater keeps going deeper until it reaches a layer of impermeable rock. When this happens, the water stops moving down. As a result, water begins filling up the pores in the rocks above. A layer of permeable rock that lets water move freely is an **aquifer** (AK wuh fur). The area where all of the pores in the rock are filled with water is the zone of saturation. The upper surface of this zone is the **water table.**

Figure 13 A stream's surface level is the water table. Below that is the zone of saturation.

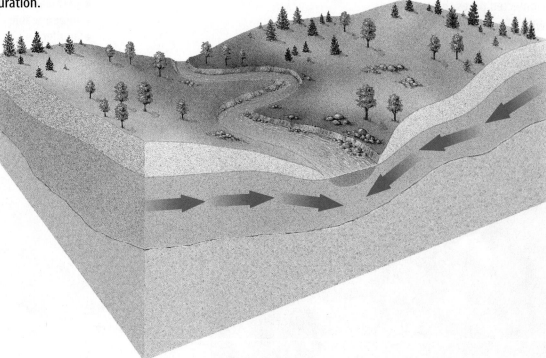

Water Table

Why are the zone of saturation and the water table so important? An average United States resident uses about 626 L of water per day. That's enough to fill nearly two thousand soft drink cans. Many people get their water from groundwater through wells that have been drilled into the zone of saturation. However, the supply of groundwater is limited. During a drought, the water table drops. This is why you should conserve water.

Applying Math — Calculate Rate of Flow

GROUNDWATER FLOW You and your family are hiking and the temperature is hot. You feel as if you can't walk one step farther. Luckily, relief is in sight. On the side of a nearby hill you see a stream, and you rush to splash some water on your face. Although you probably feel that it's taking you forever to reach the stream, your pace is quick when compared to how long it takes groundwater to flow through the aquifer that feeds the stream. The following problem will give you some idea of just how slowly groundwater flows through an aquifer.

The groundwater flows at a rate of 0.6 m/day. You've run 200 m to get some water from a stream. How long does it take the groundwater in the aquifer to travel the same distance?

Solution

1 *This is what you know:*
- the distance that the groundwater has to travel: $d = 200$ m
- the rate that groundwater flows through the aquifer: $r = 0.6$ m/day

2 *This is what you want to find:* time $= t$

3 *This is the equation you use:* $r \times t = d$ (rate $\times$ time $=$ distance)

4 *Solve the equation for* t *and then substitute known values:*

$$t = \frac{d}{r} = \frac{(200 \text{ m})}{(0.6 \text{ m/day})} = 333.33 \text{ days}$$

Practice Problems

1. The groundwater in an aquifer flows at a rate of 0.5 m/day. How far does the groundwater move in a year?

2. How long does it take groundwater in the above aquifer to move 100 m?

 For more practice, visit bookg.msscience.com/math_practice

Figure 14 The years on the pole show how much the ground level dropped in the San Joaquin Valley, California, between 1925 and 1977.

Figure 15 The pressure of water in a sloping aquifer keeps an artesian well flowing.
Describe *what limits how high water can flow in an artesian well.*

Wells A good well extends deep into the zone of saturation, past the top of the water table. Groundwater flows into the well, and a pump brings it to the surface. Because the water table sometimes drops during very dry seasons, even a good well can go dry. Then time is needed for the water table to rise, either from rainfall or through groundwater flowing from other areas of the aquifer.

Where groundwater is the main source of drinking water, the number of wells and how much water is pumped out are important. If a large factory were built in such a town, the demand on the groundwater supply would be even greater. Even in times of normal rainfall, the wells could go dry if water were taken out at a rate greater than the rate at which it can be replaced.

In areas where too much water is pumped out, the land level can sink from the weight of the sediments above the now-empty pore spaces. **Figure 14** shows what occurred when too much groundwater was removed in a region of California.

One type of well doesn't need a pump to bring water to the surface. An artesian well is a well in which water rises to the surface under pressure. Artesian wells are less common than other types of wells because of the special conditions they require.

As shown in **Figure 15,** the aquifer for an artesian well needs to be located between two impermeable layers that are sloping. Water enters at the high part of the sloping aquifer. The weight of the water in the higher part of the aquifer puts pressure on the water in the lower part. If a well is drilled into the lower part of the aquifer, the pressurized water will flow to the surface. Sometimes, the pressure is great enough to force the water into the air, forming a fountain.

☑ Reading Check *How does water move through permeable rock?*

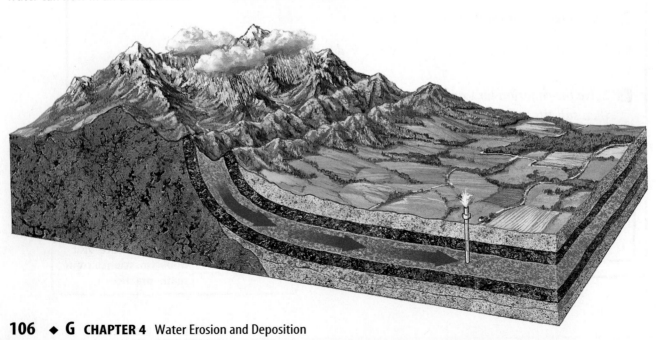

Springs In some places, the water table is so close to Earth's surface that water flows out and forms a **spring.** Springs are found on hillsides or other places where the water table meets a sloping surface. Springs often are used as a source of freshwater.

The water from most springs is a constant, cool temperature because soil and rock are good insulators and protect the groundwater from changes in temperature on Earth's surface. However, in some places, magma rises to within a few kilometers of Earth's surface and heats the surrounding rock. Groundwater that comes in contact with these hot rocks is heated and can come to the surface as a hot spring.

Geysers When water is put into a teakettle to boil, it heats slowly at first. Then some steam starts to come out of the cap on the spout, and suddenly the water starts boiling. The teakettle starts whistling as steam is forced through the cap. A similar process can occur with groundwater. One of the places where groundwater is heated is in Yellowstone National Park in Wyoming. Yellowstone has hot springs and geysers. A **geyser** is a hot spring that erupts periodically, shooting water and steam into the air. Groundwater is heated to high temperatures, causing it to expand underground. This expansion forces some of the water out of the ground, taking the pressure off the remaining water. The remaining water boils quickly, with much of it turning to steam. The steam shoots out of the opening like steam out of a teakettle, forcing the remaining water out with it. Yellowstone's famous geyser, Old Faithful, pictured in **Figure 16,** shoots between 14,000 and 32,000 L of water and steam into the air about once every 80 min.

The Work of Groundwater

Although water is the most powerful agent of erosion on Earth's surface, it also can have a great effect underground. Water mixes with carbon dioxide gas to form a weak acid called carbonic acid. Some of this carbon dioxide is absorbed from the air by rainwater or surface water. Most carbon dioxide is absorbed by groundwater moving through soil. One type of rock that is dissolved easily by this acid is limestone. Acidic groundwater moves through natural cracks and pores in limestone, dissolving the rock. Gradually, the cracks in the limestone enlarge until an underground opening called a **cave** is formed.

Acid Rain Effects Acid rain occurs when gases released by burning oil and coal mix with water in the air. Infer what effect acid rain can have on a statue made of limestone.

Figure 16 Yellowstone's famous geyser, Old Faithful, used to erupt once about every 76 min. An earthquake on January 9, 1998, slowed Old Faithful's "clock" by 4 min to an average of one eruption about every 80 min. The average height of the geyser's water is 40.5 m.

Cave Formation You've probably seen a picture of the inside of a cave, like the one shown in **Figure 17,** or perhaps you've visited one. Groundwater not only dissolves limestone to make caves, but it also can make deposits on the insides of caves.

Water often drips slowly from cracks in the cave walls and ceilings. This water contains calcium ions dissolved from the limestone. If the water evaporates while hanging from the ceiling of a cave, a deposit of calcium carbonate is left behind. Stalactites form when this happens over and over. Where drops of water fall to the floor of the cave, a stalagmite forms. The words *stalactite* and *stalagmite* come from Greek words that mean "to drip."

Sinkholes If underground rock is dissolved near the surface, a sinkhole may form. A sinkhole is a depression on the surface of the ground that forms when the roof of a cave collapses or when material near the surface dissolves. Sinkholes are common features in places like Florida and Kentucky that have lots of limestone and enough rainfall to keep the groundwater system supplied with water. Sinkholes can cause property damage if they form in a populated area.

In summary, when rain falls and becomes groundwater, it might dissolve limestone and form a cave, erupt from a geyser, or be pumped from a well to be used at your home.

Figure 17 Water dissolves rock to form caves and also deposits material to form spectacular formations, such as these in Carlsbad Caverns in New Mexico.

---

section ② review

Summary

Groundwater Systems

- Water that soaks into the ground and collects in pore spaces is called groundwater.
- 14 percent of all freshwater on Earth exists as groundwater.
- Groundwater systems have connecting pores.
- The zone of saturation is the area where all pores in the rock are filled with water.

Water Table

- The supply of groundwater is limited.

Self Check

1. **Describe** how the permeability of soil and rocks affects the flow of groundwater.
2. **Describe** why a well might go dry.
3. **Explain** how caves form.
4. **Think Critically** Why would water in wells, geysers, and hot springs contain dissolved materials?

Applying Skills

5. **Compare and contrast** wells, geysers, and hot springs.

Science Online bookg.msscience.com/self_check_quiz

Ocean Shoreline

The Shore

Picture yourself sitting on a beautiful, sandy beach like the one shown in **Figure 18.** Nearby, palm trees sway in the breeze. Children play in the quiet waves lapping at the water's edge. It's hard to imagine a place more peaceful. Now, picture yourself sitting along another shore. You're on a high cliff watching waves crash onto boulders far below. Both of these places are shorelines. An ocean shoreline is where land meets the ocean.

The two shorelines just described are different even though both experience surface waves, tides, and currents. These actions cause shorelines to change constantly. Sometimes you can see these changes from hour to hour. Why are shorelines so different? You'll understand why they look different when you learn about the forces that shape shorelines.

Figure 18 Waves, tides, and currents cause shorelines to change constantly. Waves approaching the shoreline at an angle create a longshore current.
Describe *the effects longshore currents have on a shoreline.*

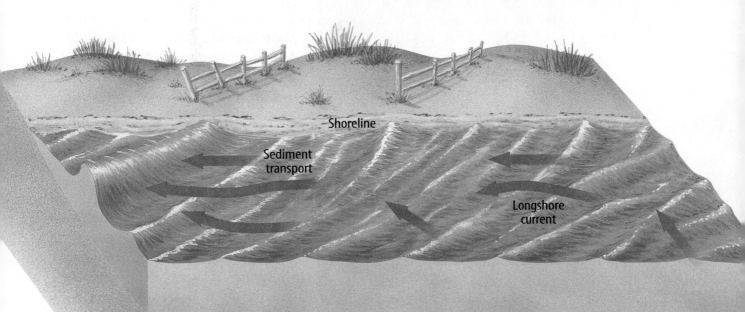

Shoreline

Sediment transport

Longshore current

Shoreline Forces When waves constantly pound against the shore, they break rocks into ever-smaller pieces. Currents move many metric tons of sediment along the shoreline. The sediment grains grind against each other like sandpaper. The tide goes out carrying sediment to deeper water. When the tide returns, it brings new sediment with it. These forces are always at work, slowly changing the shape of the shoreline. Water is always in motion along the shore.

The three major forces at work on the shoreline are waves, currents, and tides. Winds blowing across the water make waves. Waves, crashing against a shoreline, are a powerful force. They can erode and move large amounts of material in a short time. Waves usually collide with a shore at slight angles. This creates a **longshore current** of water that runs parallel to the shoreline. Longshore currents, shown in **Figure 18,** carry many metric tons of loose sediments and act like rivers of sand in the ocean.

Reading Check *How does a longshore current form?*

Tides create currents that move at right angles to the shore. These are called tidal currents. Outgoing tides carry sediments away from the shore, and incoming tides bring new sediments toward the shore. Tides work with waves to shape shorelines. You've seen the forces that affect all shorelines. Now you will see the differences that make one shore a flat, sandy beach and another shore a steep, rocky cliff.

Figure 19 Along a rocky shoreline, the force of pounding waves breaks rock fragments loose, then grinds them into smaller and smaller pieces.

Rocky Shorelines

Rocks and cliffs are the most common features along rocky shorelines like the one in **Figure 19.** Waves crash against the rocks and cliffs. Sediments in the water grind against the cliffs, slowly wearing the rock away. Then rock fragments broken from the cliffs are ground up by the endless motion of waves. They are transported as sediment by longshore currents.

Softer rocks become eroded before harder rocks do, leaving islands of harder rocks. This takes thousands of years, but remember that the ocean never stops. In a single day, about 14,000 waves crash onto shore.

This quartz sand from a Texas beach is clear and glassy.

Some Hawaiian beaches are composed of black basalt sand.

Figure 20 Beach sand varies in size, color, and composition.

Sandy Beaches

Smooth, gently sloping shorelines are different from steep, rocky shorelines. Beaches are the main feature here. **Beaches** are deposits of sediment that are parallel to the shore.

Beaches are made up of different materials. Some are made of rock fragments from the shoreline. Many beaches are made of grains of quartz, and others are made of seashell fragments. These fragments range in size from stones larger than your hand to fine sand. Sand grains range from 0.06 mm to 2 mm in diameter. Why do many beaches have particles of this size? Waves break rocks and seashells down to sand-sized particles like those shown in **Figure 20.** The constant wave motion bumps sand grains together. This bumping not only breaks particles into smaller pieces but also smooths off their jagged corners, making them more rounded.

✓ Reading Check *How do waves affect beach particles?*

Sand in some places is made of other things. For example, Hawaii's black sands are made of basalt, and its green sands are made of the mineral olivine. Jamaica's white sands are made of coral and shell fragments.

Sand Erosion and Deposition

Longshore currents carry sand along beaches to form features such as barrier islands, spits, and sandbars. Storms and wind also move sand. Thus, beaches are fragile, short-term land features that are damaged easily by storms and human activities such as some types of construction. Communities in widely separated places such as Long Island, New York; Malibu, California; and Padre Island, Texas, have problems because of beach erosion.

Barrier Islands Barrier islands are sand deposits that lay parallel to the shore but are separated from the mainland. These islands start as underwater sand ridges formed by breaking waves. Hurricanes and storms add sediment to them, raising some to sea level. When a barrier island becomes large enough, the wind blows the loose sand into dunes, keeping the new island above sea level. As with all seashore features, barrier islands are short term, lasting from a few years to a few centuries.

The forces that build barrier islands also can erode them. Storms and waves carry sediments away. Beachfront development, as in **Figure 21,** can be affected by shoreline erosion.

Figure 21 Shorelines change constantly. Human development is often at risk from shoreline erosion.

---

section 3 review

Summary

The Shore
- An ocean shoreline is where land meets the ocean.
- The forces that shape shorelines are waves, currents, and tides.

Rocky Shorelines
- Rocks and cliffs are the most common features along rocky shorelines.

Sandy Beaches
- Beaches are made up of different materials. Some are made of rock fragments and others are made of seashell fragments.

Sand Erosion and Deposition
- Longshore currents, storms, and wind move sand.
- Beaches are fragile, short-term land features.

Self Check

1. **Identify** major forces that cause shoreline erosion.
2. **Compare and contrast** the features you would find along a steep, rocky shoreline with the features you would find along a gently sloping, sandy shoreline.
3. **Explain** how the type of shoreline could affect the types of sediments you might find there.
4. **List** several materials that beach sand might be composed of. Where do these materials come from?
5. **Think Critically** How would erosion and deposition of sediment along a shoreline be affected if the longshore current was blocked by a wall built out into the water?

Applying Math

6. **Solve One-Step Equations** If 14,000 waves crash onto a shore daily, how many waves crash onto it in a year? How many crashed onto it since you were born?

Classifying Types of Sand

Sand is made of different kinds of grains, but did you realize that the slope of a beach is related to the size of its grains? The coarser the grain size is, the steeper the beach is. The composition of sand also is important. Many sands are mined because they have economic value.

▶ Real-World Question

What characteristics can be used to classify different types of beach sand?

Goals
- **Observe** differences in sand.
- **Identify** characteristics of beach sand.
- **Infer** sediment sources.

Materials
samples of different sands (3)
magnifying lens
*stereomicroscope
magnet
*Alternate materials

Safety Precautions

▶ Procedure

1. **Design** a five-column data table to compare the three sand samples. Use column one for the samples and the others for the characteristics you will be examining.

| Angular | Sub-angular | Sub-rounded | Rounded |

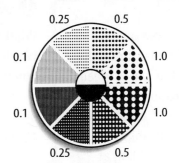

0.25   0.5
0.1   1.0
0.1   1.0
0.25   0.5

Sand gauge
(measurements in mm)

2. **Use the diagram** in the previous column to determine the average roundness of each sample.

3. **Identify** the grain size of your samples by using the sand gauge above. To determine the grain size, place sand grains in the middle of the circle of the sand gauge. Use the upper half of the circle for dark-colored particles and the bottom half for light-colored particles.

4. **Decide** on two other characteristics to examine that will help you classify your samples.

▶ Conclude and Apply

1. **Compare and contrast** some characteristics of beach sand.

2. **Describe** why there are variations in the characteristics of different sand samples.

3. **Explain** what your observations tell you about the sources of the three samples.

Communicating Your Data

Compare your results with those of other students.

Water Speed and Erosion

Goals

■ **Assemble** an apparatus for measuring the effect of water speed on erosion.

■ **Observe and measure** the ability of water traveling at different speeds to erode sand.

Materials

paint roller pan
*disposable wallpaper
    trays*
sand
1-L beaker
rubber tubing (20 cm)
metric ruler
water
stopwatch
fine-mesh screen
wood block
Alternate materials

Safety Precautions

Wash your hands after you handle the sand. Immediately clean up any water that spills on the floor.

◉ Real-World Question

What would it be like to make a raft and use it to float on a river? Would it be easy? Would you feel like Tom Sawyer? Probably not. You'd be at the mercy of the current. Strong currents create fast rivers. But does fast moving water affect more than just floating rafts and other objects? How does the speed of a stream or river affect its ability to erode?

◉ Procedure

1. Copy the data table on the following page.

2. Place the screen in the sink. Pour moist sand into your pan and smooth out the sand. Set one end of the pan on the wood block and hang the other end over the screen in the sink. Excess water will flow onto the screen in the sink.

3. Attach one end of the hose to the faucet and place the other end in the beaker. Turn on the water so that it trickles into the beaker. Time how long it takes for the trickle of water to fill the beaker to the 1-L mark. Divide 1 L by your time in seconds to calculate the water speed. Record the speed in your data table.

4. Without altering the water speed, hold the hose over the end of the pan that is resting on the wood block. Allow the water to flow into the sand for 2 min. At the end of 2 min, turn off the water.

5. **Measure** the depth and length of the eroded channel formed by the water. Count the number of branches formed on the channel. Record your measurements and observations in your data table.

Water Speed and Erosion			
Water Speed (Liters per Second)	Depth of Channel	Length of Channel	Number of Channel Branches
	Do not write in this book.		

6. Empty the excess water from the tray and smooth out the sand. Repeat steps 3 through 5 two more times, increasing your water speed each time.

⦾ Conclude and Apply

1. **Identify** the constants and variables in your experiment.

2. **Observe** Which water speed created the deepest and longest channel?

3. **Observe** Which water speed created the greatest number of branches?

4. **Infer** the effect that water speed has on erosion.

5. **Predict** how your results would have differed if one end of the pan had been raised higher.

6. **Infer** how streams and rivers can shape Earth's surface.

𝒞ommunicating Your Data

Write a pamphlet for people buying homes near rivers or streams that outlines the different effects that water erosion could have on their property.

Is there hope for America's coastlines or is beach erosion a "shore" thing?

Sands in Time

Water levels are rising along the coastline of the United States. Serious storms and the building of homes and businesses along the shore are leading to the erosion of anywhere from 70 percent to 90 percent of the U.S. coastline. A report from the Federal Emergency Management Agency (FEMA) confirms this. The report says that one meter of United States beaches will be eaten away each year for the next 60 years. Since 1965, the federal government has spent millions of dollars replenishing more than 1,300 eroding sandy shores around the country. And still beaches continue to disappear.

The slowly eroding beaches are upsetting to residents and officials of many communities, who depend on their shore to earn money from visitors. Some city and state governments are turning to beach nourishment—a process in which sand is taken from the seafloor and dumped on beaches. The process is expensive, however. The state of Delaware, for example, is spending 7,000,000 dollars to bring in sand for its beaches.

This beach house will collapse as its underpinnings are eroded.

Other methods of saving eroding beaches are being tried. In places along the Great Lakes shores and coastal shores, one company has installed fabrics underwater to slow currents. By slowing currents, sand is naturally deposited and kept in place.

Another shore-saving device is a synthetic barrier that is shaped like a plastic snowflake. A string of these barriers is secured just offshore. They absorb the energy of incoming waves. Reducing wave energy can prevent sand from being eroded from the beach. New sand also might accumulate because the barriers slow down the currents that flow along the shore.

Many people believe that communities along the shore must restrict the beachfront building of homes, hotels, and stores. Since some estimates claim that by the year 2025, nearly 75 percent of the U.S. population will live in coastal areas, it's a tough solution. Says one geologist, "We can retreat now and save our beaches or we can retreat later and probably ruin the beaches in the process."

Debate Using the facts in this article and other research you have done in your school media center or through Web links at msscience.com, make a list of methods that could be used to save beaches. Debate the issue with your classmates.

Science online

For more information, visit bookg.msscience.com/time

Reviewing Main Ideas

Section 1 Surface Water

1. Rainwater that does not soak into the ground is pulled down the slope by gravity. This water is called runoff.

2. Runoff can erode sediment. Factors such as steepness of slope and number and type of plants affect the amount of erosion. Rill, gully, and sheet erosion are types of surface water erosion caused by runoff.

3. Runoff generally flows into streams that merge with larger rivers until emptying into a lake or ocean. Major river systems usually contain several different types of streams.

4. Young streams flow through steep valleys and have rapids and waterfalls. Mature streams flow through gentler terrain and have less energy. Old streams are often wide and meander across their floodplains.

Section 2 Groundwater

1. When water soaks into the ground, it becomes part of a vast groundwater system.

2. Although rock may seem solid, many types are filled with connected spaces called pores. Such rocks are permeable and can contain large amounts of groundwater.

Section 3 Ocean Shoreline

1. Ocean shorelines are always changing.

2. Waves and currents have tremendous amounts of energy which break up rocks into tiny fragments called sediment. Over time, the deposition and relocation of sediment can change beaches, sandbars, and barrier islands.

Visualizing Main Ideas

Copy and complete the following concept map on caves.

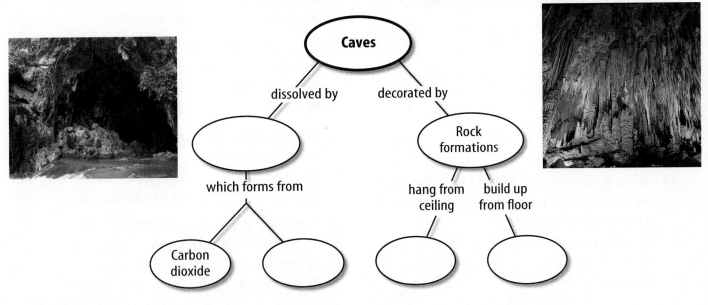

Using Vocabulary

aquifer p. 104
beach p. 111
cave p. 107
channel p. 94
drainage basin p. 96
geyser p. 107
groundwater p. 103
impermeable p. 104

longshore current p. 110
meander p. 97
permeable p. 104
runoff p. 92
sheet erosion p. 95
spring p. 107
water table p. 104

Explain the difference between the vocabulary words in each of the following sets.

1. runoff—sheet erosion
2. channel—drainage basin
3. aquifer—cave
4. spring—geyser
5. permeable—impermeable
6. sheet erosion—meander
7. groundwater—water table
8. permeable—aquifer
9. longshore current—beach
10. meander—channel

Checking Concepts

Choose the word or phrase that best answers the question.

11. Where are beaches most common?
    A) rocky shorelines
    B) flat shorelines
    C) aquifers
    D) young streams

12. What is the network formed by a river and all the smaller streams that contribute to it?
    A) groundwater system
    B) zone of saturation
    C) river system
    D) water table

13. Why does water rise in an artesian well?
    A) a pump   **C)** heat
    B) erosion   **D)** pressure

14. Which term describes rock through which fluids can flow easily?
    A) impermeable   **C)** saturated
    B) meanders   **D)** permeable

15. Identify an example of a structure created by deposition.
    A) beach   **C)** cave
    B) rill   **D)** geyser

16. Which stage of development are mountain streams in?
    A) young   **C)** old
    B) mature   **D)** meandering

17. What forms as a result of the water table meeting Earth's surface?
    A) meander   **C)** aquifer
    B) spring   **D)** stalactite

18. What contains heated groundwater that reaches Earth's surface?
    A) water table   **C)** aquifer
    B) cave   **D)** hot spring

19. What is a layer of permeable rock that water flows through?
    A) an aquifer   **C)** a water table
    B) a pore   **D)** impermeable

20. Name the deposit that forms when a mountain river runs onto a plain.
    A) subsidence   **C)** infiltration
    B) an alluvial fan **D)** water diversion

Science Online bookg.msscience.com/vocabulary_puzzlemaker

Thinking Critically

21. Concept Map Copy and complete the concept map below using the following terms: *developed meanders, gentle curves, gentle gradient, old, rapids, steep gradient, wide floodplain,* and *young*.

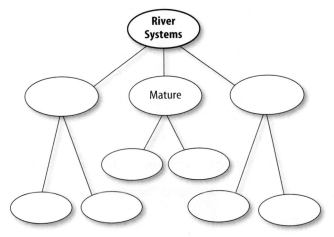

22. Describe what determines whether a stream erodes its bottom or its sides.

23. Interpret Data The rate of water flowing out of the Brahmaputra River in India, the La Plata River in South America, and the Mississippi River in North America are given in the table below. Infer which river carries the most sediment.

River Flow Rates

River	Flow (m³/s)
Brahmaputra River, India	19,800
La Plata River, South America	79,300
Mississippi River, North America	175,000

24. Explain why the Mississippi River has meanders along its course.

25. Outline Make an outline that explains the three stages of stream development.

26. Form Hypotheses Hypothesize why most of the silt in the Mississippi delta is found farther out to sea than the sand-sized particles are.

27. Infer Along what kind of shoreline would you find barrier islands?

28. Explain why you might be concerned if developers of a new housing project started drilling wells near your well.

29. Use Variables, Constants, and Controls Explain how you could test the effect of slope on the amount of runoff produced.

Performance Activities

30. Poster Research a beach that interests you. Make a poster that shows different features you would find at a beach.

Applying Math

Use the illustration below to answer questions 31–32.

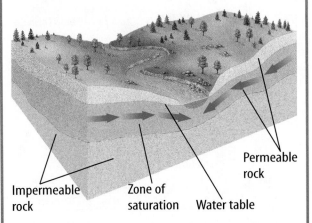

Impermeable rock   Zone of saturation   Water table   Permeable rock

31. Flow Distance The groundwater in an aquifer flows at a rate of 0.2 m/day. How far does the groundwater move in one week?

32. Flow Time If groundwater in an aquifer flows at a rate of 0.4 m/day, how long does it take groundwater to move 24 m?

Part 1 | Multiple Choice

Record your answers on the answer sheet provided by your teacher or on a sheet of paper.

1. Which is erosion over a large, flat area?
   - **A.** gully
   - **B.** rill
   - **C.** runoff
   - **D.** sheet

2. Which is erosion where a rill becomes broader and deeper?
   - **A.** gully
   - **B.** rill
   - **C.** runoff
   - **D.** sheet

3. Which is the area of land from which a stream collects runoff?
   - **A.** drainage basin
   - **B.** gully
   - **C.** runoff
   - **D.** stream channel

4. Which type of soil or rock allows water to pass through them?
   - **A.** impermeable
   - **B.** nonporous
   - **C.** permeable
   - **D.** underground

Refer to the figure below to answer question 5.

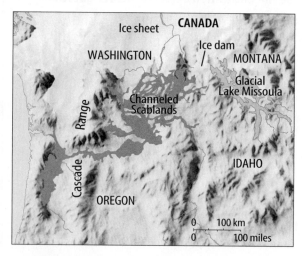

5. Which formed the Channeled Scablands?
   - **A.** deposition
   - **B.** floodwaters
   - **C.** rill erosion
   - **D.** sheet erosion

6. Which dissolves limestone to form caves?
   - **A.** carbonic acid
   - **B.** hydrochloric acid
   - **C.** stalactites
   - **D.** stalagmites

7. Which forms on the ceilings of caves as water drips through cracks?
   - **A.** aquifer
   - **B.** geyser
   - **C.** stalactite
   - **D.** stalagmite

8. Which are piles of sand found on barrier islands?
   - **A.** deltas
   - **B.** dunes
   - **C.** geysers
   - **D.** streams

9. Which creates springs and geysers?
   - **A.** groundwater
   - **B.** gullies
   - **C.** rills
   - **D.** runoff

Refer to the figure below to answer question 10 and 11.

10. Which feature is shown?
    - **A.** artesian well
    - **B.** aquifer
    - **C.** geyser
    - **D.** waterfall

11. Which provides the water?
    - **A.** groundwater
    - **B.** runoff
    - **C.** stream
    - **D.** surface water

12. Which is a layer of permeable rock through which water moves freely?
    - **A.** aquifer
    - **B.** clay
    - **C.** geyser
    - **D.** granite

Test-Taking Tip

Correct Answer Bubbles For each question, double check that you are filling in the correct answer bubble for the question number you are working on.

Part 2 | Short Response/Grid In

Record your answers on the answer sheet provided by your teacher or on a sheet of paper.

13. How does gravity affect water erosion?

14. Describe the different types of load in a stream.

15. What will happen to homes and businesses located in a floodplain?

16. Explain how the water table and the zone of saturation are related.

17. How would harmful chemicals in the soil enter into the groundwater system?

Refer to the picture below to answer questions 18 and 19.

18. What type of erosion is shown?

19. What led to the erosion shown?

20. What factors affect runoff?

21. What are the characteristics of a mature stream?

Part 3 | Open Ended

Record your answers on a sheet of paper.

22. Compare and contrast rocky shorelines and sandy beaches.

23. What are the Outer Banks in North Carolina and how were they formed?

24. What can humans do to try to control flood waters?

25. Explain how waves produce longshore currents.

Refer to the figure below to answer question 26.

26. What stream formation is shown in the diagram above? Explain how differing speeds of water in a stream can cause this type of stream formation.

27. Explain how a drainage basin works. Compare and contrast a drainage basin to the gutter drainage system of a rooftop.

28. Compare and contrast a pumped well and an artesian well.

29. Compare and contrast alluvial fans and deltas.

Clues to Earth's Past

Reading the Past

The pages of Earth's history, much like the pages of human history, can be read if you look in the right place. Unlike the pages of a book, the pages of Earth's past are written in stone. In this chapter you will learn how to read the pages of Earth's history to understand what the planet was like in the distant past.

Science Journal List three fossils that you would expect to find a million years from now in the place you live today.

Start-Up Activities

Clues to Life's Past

Fossil formation begins when dead plants or animals are buried in sediment. In time, if conditions are right, the sediment hardens into sedimentary rock. Parts of the organism are preserved along with the impressions of parts that don't survive. Any evidence of once-living things contained in the rock record is a fossil.

1. Fill a small jar (about 500 mL) one-third full of plaster of paris. Add water until the jar is half full.

2. Drop in a few small shells.

3. Cover the jar and shake it to model a swift, muddy stream.

4. Now model the stream flowing into a lake by uncovering the jar and pouring the contents into a paper or plastic bowl. Let the mixture sit for an hour.

5. Crack open the hardened plaster to locate the model fossils.

6. **Think Critically** Remove the shells from the plaster and study the impressions they made. In your Science Journal, list what the impressions would tell you if found in a rock.

Age of Rocks Make the following Foldable to help you understand how scientists determine the age of a rock.

STEP 1 Fold a sheet of paper in half lengthwise.

STEP 2 Fold paper down 2.5 cm from the top. (Hint: From the tip of your index finger to your middle knuckle is about 2.5 cm.)

STEP 3 Open and draw lines along the 2.5-cm fold. Label as shown.

Summarize in a Table As you read the chapter, in the left column, list four different ways in which one could determine the age of a rock. In the right column, note whether each method gives an absolute or a relative age.

Preview this chapter's content and activities at
bookg.msscience.com

Get Ready to Read

Take Notes

1 Learn It! The best way for you to remember information is to write it down, or take notes. Good note-taking is useful for studying and research. When you are taking notes, it is helpful to
- phrase the information in your own words;
- restate ideas in short, memorable phrases;
- stay focused on main ideas and only the most important supporting details.

2 Practice It! Make note-taking easier by using a chart to help you organize information clearly. Write the main ideas in the left column. Then write at least three supporting details in the right column. Read the text from Section 1 of this chapter under the heading *Conditions Needed for Fossil Formation,* page 125. Then take notes using a chart, such as the one below.

Main Idea	Supporting Details
	1. 2. 3. 4. 5.
	1. 2. 3. 4. 5.

3 Apply It! As you read this chapter, make a chart of the main ideas. Next to each main idea, list at least three supporting details.

Target Your Reading

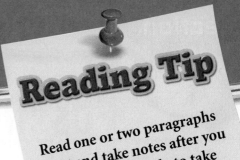

Reading Tip

Read one or two paragraphs first and take notes after you read. You are likely to take down too much information if you take notes as you read.

Use this to focus on the main ideas as you read the chapter.

1 Before you read the chapter, respond to the statements below on your worksheet or on a numbered sheet of paper.

- Write an **A** if you **agree** with the statement.
- Write a **D** if you **disagree** with the statement.

2 After you read the chapter, look back to this page to see if you've changed your mind about any of the statements.

- If any of your answers changed, explain why.
- Change any false statements into true statements.
- Use your revised statements as a study guide.

Science Online

Print out a worksheet of this page at bookg.msscience.com

Before You Read A or D		Statement	After You Read A or D
	1	All fossils are made from the hard parts of animals.	
	2	Fossils can be used as evidence to show that past climates and environments have changed.	
	3	A trace fossil is the outline, or copy, of a fossil.	
	4	Sediment typically accumulates in horizontal beds, which can later form layers of sedimentary rock.	
	5	The relative age of a rock layer indicates whether the layer is older or younger when compared to other rock layers.	
	6	The principle of superposition refers to a high concentration of fossils within a small area.	
	7	Most sequences of rock layers are complete.	
	8	Geologists often can match up, or correlate, layers of rock over great distances.	
	9	The absolute age of a material refers to the actual age, in years, of the material.	

Fossils

What You'll Learn

- **List** the conditions necessary for fossils to form.
- **Describe** several processes of fossil formation.
- **Explain** how fossil correlation is used to determine rock ages.
- **Determine** how fossils can be used to explain changes in Earth's surface, life forms, and environments.

Why It's Important

Fossils help scientists find oil and other sources of energy necessary for society.

Review Vocabulary

paleontologist: a scientist who studies fossils

New Vocabulary

- fossil
- permineralized remains
- carbon film
- mold
- cast
- index fossil

Traces of the Distant Past

A giant crocodile lurks in the shallow water of a river. A herd of *Triceratops* emerges from the edge of the forest and cautiously moves toward the river. The dinosaurs are thirsty, but danger waits for them in the water. A large bull *Triceratops* moves into the river. The others follow.

Does this scene sound familiar to you? It's likely that you've read about dinosaurs and other past inhabitants of Earth. But how do you know that they really existed or what they were like? What evidence do humans have of past life on Earth? The answer is fossils. Paleontologists, scientists who study fossils, can learn about extinct animals from their fossil remains, as shown in **Figure 1.**

Figure 1 Scientists can learn how dinosaurs looked and moved using fossil remains. A skeleton can then be reassembled and displayed in a museum.

Formation of Fossils

Fossils are the remains, imprints, or traces of prehistoric organisms. Fossils have helped scientists determine approximately when life first appeared, when plants and animals first lived on land, and when organisms became extinct. Fossils are evidence of not only when and where organisms once lived, but also how they lived.

For the most part, the remains of dead plants and animals disappear quickly. Scavengers eat and scatter the remains of dead organisms. Fungi and bacteria invade, causing the remains to rot and disappear. If you've ever left a banana on the counter too long, you've seen this process begin. In time, compounds within the banana cause it to break down chemically and soften. Microorganisms, such as bacteria, cause it to decay. What keeps some plants and animals from disappearing before they become fossils? Which organisms are more likely to become fossils?

Conditions Needed for Fossil Formation Whether or not a dead organism becomes a fossil depends upon how well it is protected from scavengers and agents of physical destruction, such as waves and currents. One way a dead organism can be protected is for sediment to bury the body quickly. If a fish dies and sinks to the bottom of a lake, sediment carried into the lake by a stream can cover the fish rapidly. As a result, no waves or scavengers can get to it and tear it apart. The body parts then might be fossilized and included in a sedimentary rock like shale. However, quick burial alone isn't always enough to make a fossil.

Organisms have a better chance of becoming fossils if they have hard parts such as bones, shells, or teeth. One reason is that scavengers are less likely to eat these hard parts. Hard parts also decay more slowly than soft parts do. Most fossils are the hard parts of organisms, such as the fossil teeth in **Figure 2.**

Types of Preservation

Perhaps you've seen skeletal remains of *Tyrannosaurus rex* towering above you in a museum. You also have some idea of what this dinosaur looked like because you've seen illustrations. Artists who draw *Tyrannosaurus rex* and other dinosaurs base their illustrations on fossil bones. What preserves fossil bones?

Figure 2 These fossil shark teeth are hard parts. Soft parts of animals do not become fossilized as easily.

Predicting Fossil Preservation

Procedure
1. Take a brief walk outside and observe your neighborhood.
2. Look around and notice what kinds of plants and animals live nearby.

Analysis
1. Predict what remains from your time might be preserved far into the future.
2. Explain what conditions would need to exist for these remains to be fossilized.

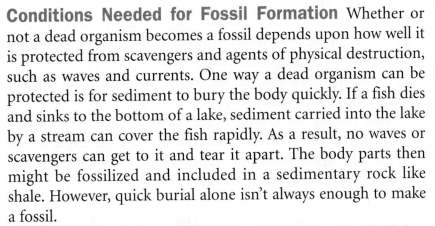

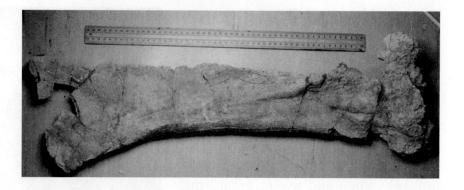

Figure 3 Opal and various minerals have replaced original materials and filled the hollow spaces in this permineralized dinosaur bone. **Explain** *why this fossil retained the shape of the original bone.*

Mineral Replacement Most hard parts of organisms such as bones, teeth, and shells have tiny spaces within them. In life, these spaces can be filled with cells, blood vessels, nerves, or air. When the organism dies and the soft materials inside the hard parts decay, the tiny spaces become empty. If the hard part is buried, groundwater can seep in and deposit minerals in the spaces. **Permineralized remains** are fossils in which the spaces inside are filled with minerals from groundwater. In permineralized remains, some original material from the fossil organism's body might be preserved—encased within the minerals from groundwater. It is from these original materials that DNA, the chemical that contains an organism's genetic code, can sometimes be recovered.

Sometimes minerals replace the hard parts of fossil organisms. For example, a solution of water and dissolved silica (the compound SiO_2) might flow into and through the shell of a dead organism. If the water dissolves the shell and leaves silica in its place, the original shell is replaced.

Often people learn about past forms of life from bones, wood, and other remains that became permineralized or replaced with minerals from groundwater, as shown in **Figure 3,** but many other types of fossils can be found.

Figure 4 Graptolites lived hundreds of millions of years ago and drifted on currents in the oceans. These organisms often are preserved as carbon films.

Carbon Films The tissues of organisms are made of compounds that contain carbon. Sometimes fossils contain only carbon. Fossils usually form when sediments bury a dead organism. As sediment piles up, the organism's remains are subjected to pressure and heat. These conditions force gases and liquids from the body. A thin film of carbon residue is left, forming a silhouette of the original organism called a **carbon film. Figure 4** shows the carbonized remains of graptolites, which were small marine animals. Graptolites have been found in rocks as old as 500 million years.

Coal In swampy regions, large volumes of plant matter accumulate. Over millions of years, these deposits become completely carbonized, forming coal. Coal is an important fuel source, but since the structure of the original plant is usually lost, it cannot reveal as much about the past as other kinds of fossils.

Reading Check *In what sort of environment does coal form?*

Molds and Casts In nature, impressions form when seashells or other hard parts of organisms fall into a soft sediment such as mud. The object and sediment are then buried by more sediment. Compaction, together with cementation, which is the deposition of minerals from water into the pore spaces between sediment particles, turns the sediment into rock. Other open pores in the rock then let water and air reach the shell or hard part. The hard part might decay or dissolve, leaving behind a cavity in the rock called a **mold.** Later, mineral-rich water or other sediment might enter the cavity, form new rock, and produce a copy or **cast** of the original object, as shown in **Figure 5.**

INTEGRATE Social Studies

Coal Mining Many of the first coal mines in the United States were located in eastern states like Pennsylvania and West Virginia. In your Science Journal, discuss how the environments of the past relate to people's lives today.

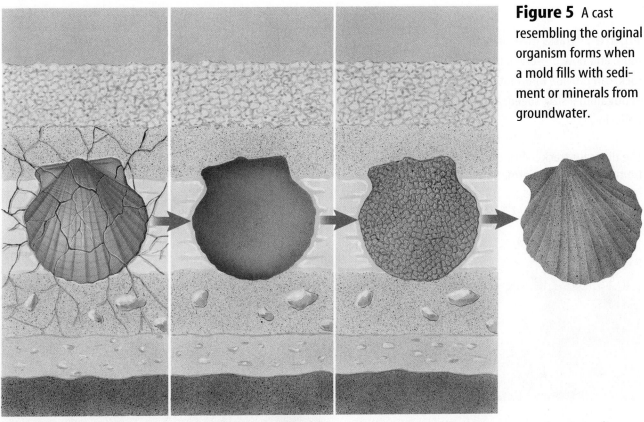

Figure 5 A cast resembling the original organism forms when a mold fills with sediment or minerals from groundwater.

The fossil begins to dissolve as water moves through spaces in the rock layers.

The fossil has been dissolved away. The harder rock once surrounding it forms a mold.

Sediment washes into the mold and is deposited, or mineral crystals form.

A cast results.

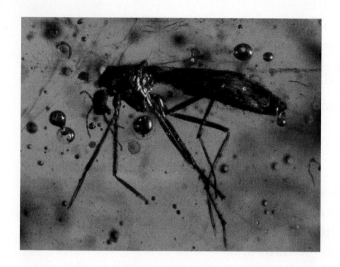

Original Remains Sometimes conditions allow original soft parts of organisms to be preserved for thousands or millions of years. For example, insects can be trapped in amber, a hardened form of sticky tree resin. The amber surrounds and protects the original material of the insect's exoskeleton from destruction, as shown in **Figure 6.** Some organisms, such as the mammoth, have been found preserved in frozen ground in Siberia. Original remains also have been found in natural tar deposits, such as the La Brea tar pits in California.

Figure 6 The original soft parts of this mosquito have been preserved in amber for millions of years.

Trace Fossils Do you have a handprint in plaster that you made when you were in kindergarten? If so, it's a record that tells something about you. From it, others can guess your size and maybe your weight at that age. Animals walking on Earth long ago left similar tracks, such as those in **Figure 7.** Trace fossils are fossilized tracks and other evidence of the activity of organisms. In some cases, tracks can tell you more about how an organism lived than any other type of fossil. For example, from a set of tracks at Davenport Ranch, Texas, you might be able to learn something about the social life of sauropods, which were large, plant-eating dinosaurs. The largest tracks of the herd are on the outer edges and the smallest are on the inside. These tracks led some scientists to hypothesize that adult sauropods surrounded their young as they traveled—perhaps to protect them from predators. A nearby set of tracks might mean that another type of dinosaur, an allosaur, was stalking the herd.

Figure 7 Tracks made in soft mud, and now preserved in solid rock, can provide information about animal size, speed, and behavior.

The dinosaur track below is from the Glen Rose Formation in north-central Texas.

The tracks to the right are located on a Navajo reservation in Arizona.

Trails and Burrows Other trace fossils include trails and burrows made by worms and other animals. These, too, tell something about how these animals lived. For example, by examining fossil burrows you can sometimes tell how firm the sediment the animals lived in was. As you can see, fossils can tell a great deal about the organisms that have inhabited Earth.

✔ **Reading Check** *How are trace fossils different from fossils that are the remains of an organism's body?*

Index Fossils

One thing you can learn by studying fossils is that species of organisms have changed over time. Some species of organisms inhabited Earth for long periods of time without changing. Other species changed a lot in comparatively short amounts of time. It is these organisms that scientists use as index fossils.

Index fossils are the remains of species that existed on Earth for relatively short periods of time, were abundant, and were widespread geographically. Because the organisms that became index fossils lived only during specific intervals of geologic time, geologists can estimate the ages of rock layers based on the particular index fossils they contain. However, not all rocks contain index fossils. Another way to approximate the age of a rock layer is to compare the spans of time, or ranges, over which more than one fossil appears. The estimated age is the time interval where fossil ranges overlap, as shown in **Figure 8.**

Figure 8 The fossils in a sequence of sedimentary rock can be used to estimate the ages of each layer. The chart shows when each organism inhabited Earth.
Explain *why it is possible to say that the middle layer of rock was deposited between 440 million and 410 million years ago.*

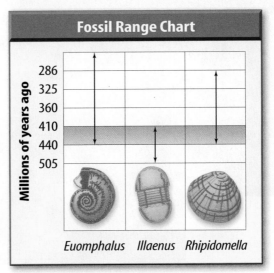

Fossil Range Chart

Millions of years ago
286
325
360
410
440
505

Euomphalus *Illaenus* *Rhipidomella*

INTEGRATE Life Science

Ancient Ecology
Ecology is the study of how organisms interact with each other and with their environment. Some paleontologists study the ecology of ancient organisms. Discuss the kinds of information you could use to determine how ancient organisms interacted with their environment.

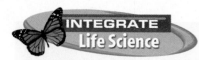

Fossils and Ancient Environments

Scientists can use fossils to determine what the environment of an area was like long ago. Using fossils, you might be able to find out whether an area was land or whether it was covered by an ocean at a particular time. If the region was covered by ocean, it might even be possible to learn the depth of the water. What clues about the depth of water do you think fossils could provide?

Fossils also are used to determine the past climate of a region. For example, rocks in parts of the eastern United States contain fossils of tropical plants. The environment of this part of the United States today isn't tropical. However, because of the fossils, scientists know that it was tropical when these plants were living. **Figure 9** shows that North America was located near the equator when these fossils formed.

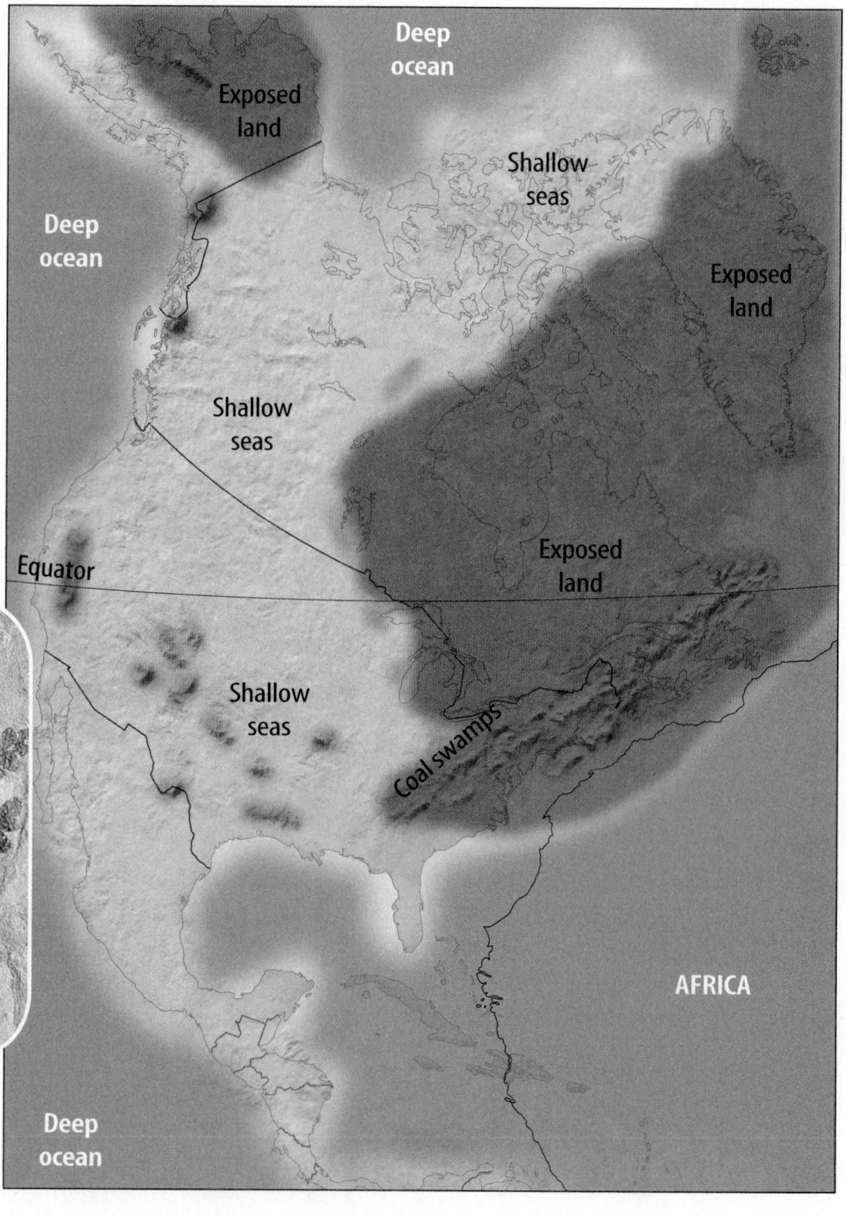

Figure 9 The equator passed through North America 310 million years ago. At this time, warm, shallow seas and coal swamps covered much of the continent, and ferns like the *Neuropteris,* below, were common.

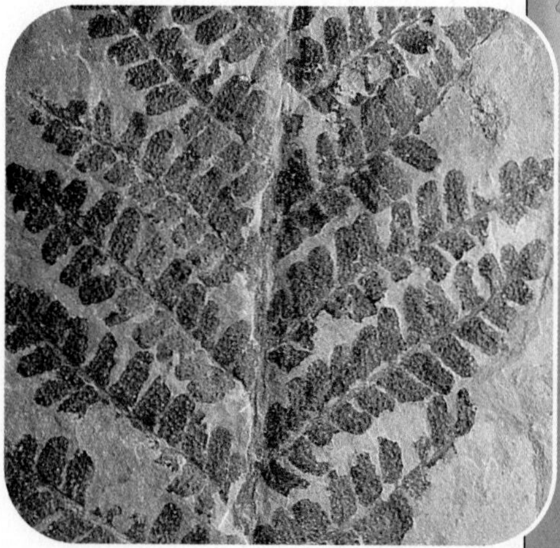

Shallow Seas How would you explain the presence of fossilized crinoids—animals that lived in shallow seas—in rocks found in what is today a desert? **Figure 10** shows a fossil crinoid and a living crinoid. When the fossil crinoids were alive, a shallow sea covered much of western and central North America. The crinoid hard parts were included in rocks that formed from the sediments at the bottom of this sea. Fossils provide information about past life on Earth and also about the history of the rock layers that contain them. Fossils can provide information about the ages of rocks and the climate and type of environment that existed when the rocks formed.

Figure 10 The crinoid on the left lived in warm, shallow seas that once covered part of North America. Crinoids like the one on the right typically live in warm, shallow waters in the Pacific Ocean.

section 1 review

Summary

Formation of Fossils

- Fossils are the remains, imprints, or traces of past organisms.
- Fossilization is most likely if the organism had hard parts and was buried quickly.

Fossil Preservation

- Permineralized remains have open spaces filled with minerals from groundwater.
- Thin carbon films remain in the shapes of dead organisms.
- Hard parts dissolve to leave molds.
- Trace fossils are evidence of past activity.

Index Fossils

- Index fossils are from species that were abundant briefly, but over wide areas.
- Scientists can estimate the ages of rocks containing index fossils.

Fossils and Ancient Environments

- Fossils tell us about the environment in which the organisms lived.

Self Check

1. **Describe** the typical conditions necessary for fossil formation.
2. **Explain** how a fossil mold is different from a fossil cast.
3. **Discuss** how the characteristics of an index fossil are useful to geologists.
4. **Describe** how carbon films form.
5. **Think Critically** What can you say about the ages of two widely separated layers of rock that contain the same type of fossil?

Applying Skills

6. **Communicate** what you learn about fossils. Visit a museum that has fossils on display. Make an illustration of each fossil in your Science Journal. Write a brief description, noting key facts about each fossil and how each fossil might have formed.
7. **Compare and contrast** original remains with other kinds of fossils. What kinds of information would only be available from original remains? Are there any limitations to the use of original remains?

Relative Ages of Rocks

as you read

What You'll Learn

- **Describe** methods used to assign relative ages to rock layers.
- **Interpret** gaps in the rock record.
- **Give** an example of how rock layers can be correlated with other rock layers.

Why It's Important

Being able to determine the age of rock layers is important in trying to understand a history of Earth.

⚙ Review Vocabulary

sedimentary rock: rock formed when sediments are cemented and compacted or when minerals are precipitated from solution

New Vocabulary

- principle of superposition
- relative age
- unconformity

Superposition

Imagine that you are walking to your favorite store and you happen to notice an interesting car go by. You're not sure what kind it is, but you remember that you read an article about it. You decide to look it up. At home you have a stack of magazines from the past year, as seen in **Figure 11.**

You know that the article you're thinking of came out in the January edition, so it must be near the bottom of the pile. As you dig downward, you find magazines from March, then February. January must be next. How did you know that the January issue of the magazine would be on the bottom? To find the older edition under newer ones, you applied the principle of superposition.

Oldest Rocks on the Bottom According to the **principle of superposition,** in undisturbed layers of rock, the oldest rocks are on the bottom and the rocks become progressively younger toward the top. Why is this the case?

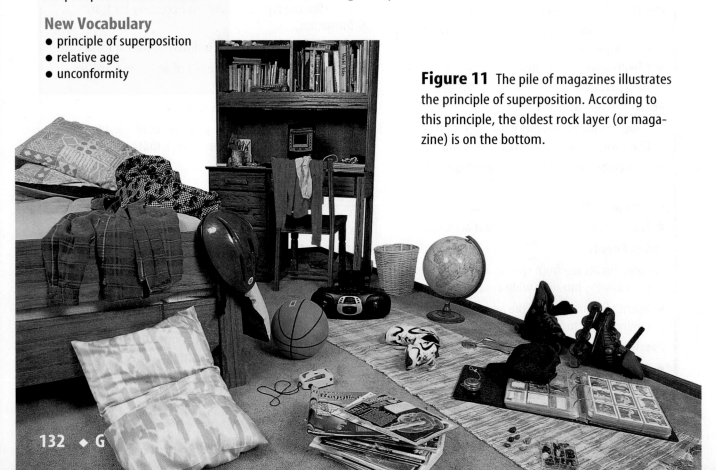

Figure 11 The pile of magazines illustrates the principle of superposition. According to this principle, the oldest rock layer (or magazine) is on the bottom.

Rock Layers Sediment accumulates in horizontal beds, forming layers of sedimentary rock. The first layer to form is on the bottom. The next layer forms on top of the previous one. Because of this, the oldest rocks are at the bottom. However, forces generated by mountain formation sometimes can turn layers over. When layers have been turned upside down, it's necessary to use other clues in the rock layers to determine their original positions and relative ages.

Relative Ages

Now you want to look for another magazine. You're not sure how old it is, but you know it arrived after the January issue. You can find it in the stack by using the principle of relative age.

The **relative age** of something is its age in comparison to the ages of other things. Geologists determine the relative ages of rocks and other structures by examining their places in a sequence. For example, if layers of sedimentary rock are offset by a fault, which is a break in Earth's surface, you know that the layers had to be there before a fault could cut through them. The relative age of the rocks is older than the relative age of the fault. Relative age determination doesn't tell you anything about the age of rock layers in actual years. You don't know if a layer is 100 million or 10,000 years old. You only know that it's younger than the layers below it and older than the fault cutting through it.

Other Clues Help Determination of relative age is easy if the rocks haven't been faulted or turned upside down. For example, look at **Figure 12.** Which layer is the oldest? In cases where rock layers have been disturbed you might have to look for fossils and other clues to date the rocks. If you find a fossil in the top layer that's older than a fossil in a lower layer, you can hypothesize that layers have been turned upside down by folding during mountain building.

Science nline

Topic: Relative Dating
Visit bookg.msscience.com for Web links to information about relative dating of rocks and other materials.

Activity Imagine yourself at an archaeological dig. You have found a rare artifact and want to know its age. Make a list of clues you might look for to provide a relative date and explain how each would allow you to approximate the artifact's age.

Figure 12 In a stack of undisturbed sedimentary rocks, the oldest rocks are at the bottom. This stack of rocks can be folded by forces within Earth.
Explain *how you can tell if an older rock is above a younger one.*

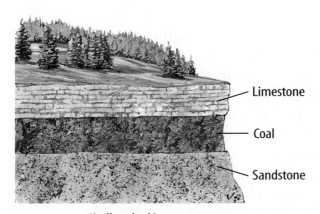

Undisturbed Layers

Limestone

Coal

Sandstone

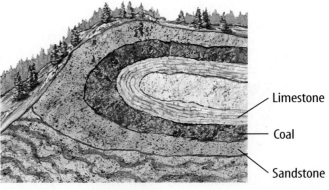

Folded Layers

Limestone

Coal

Sandstone

Figure 13 An angular unconformity results when horizontal layers cover tilted, eroded layers.

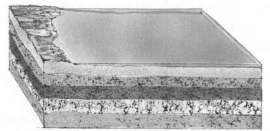

A Sedimentary rocks are deposited originally as horizontal layers.

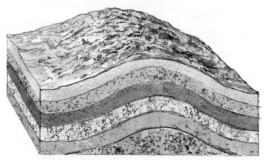

B The horizontal rock layers are tilted as forces within Earth deform them.

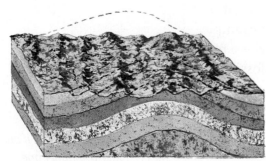

C The tilted layers erode.

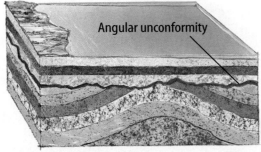

Angular unconformity

D An angular unconformity results when new layers form on the tilted layers as deposition resumes.

Unconformities

A sequence of rock is a record of past events. But most rock sequences are incomplete—layers are missing. These gaps in rock sequences are called **unconformities** (un kun FOR muh teez). Unconformities develop when agents of erosion such as running water or glaciers remove rock layers by washing or scraping them away.

✔ Reading Check *How do unconformities form?*

Angular Unconformities Horizontal layers of sedimentary rock often are tilted and uplifted. Erosion and weathering then wear down these tilted rock layers. Eventually, younger sediment layers are deposited horizontally on top of the tilted and eroded layers. Geologists call such an unconformity an angular unconformity. **Figure 13** shows how angular unconformities develop.

Disconformity Suppose you're looking at a stack of sedimentary rock layers. They look complete, but layers are missing. If you look closely, you might find an old surface of erosion. This records a time when the rocks were exposed and eroded. Later, younger rocks formed above the erosion surface when deposition of sediment began again. Even though all the layers are parallel, the rock record still has a gap. This type of unconformity is called a disconformity. A disconformity also forms when a period of time passes without any new deposition occurring to form new layers of rock.

Nonconformity Another type of unconformity, called a nonconformity, occurs when metamorphic or igneous rocks are uplifted and eroded. Sedimentary rocks are then deposited on top of this erosion surface. The surface between the two rock types is a nonconformity. Sometimes rock fragments from below are incorporated into sediments deposited above the nonconformity. All types of unconformities are shown in **Figure 14.**

Figure 14

An unconformity is a gap in the rock record caused by erosion or a pause in deposition. There are three major kinds of unconformities—nonconformity, angular unconformity, and disconformity.

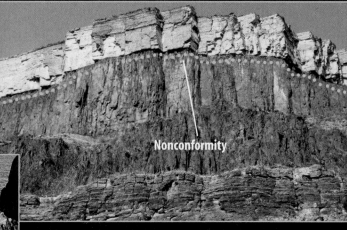

▲ In a nonconformity, horizontal layers of sedimentary rock overlie older igneous or metamorphic rocks. A nonconformity in Big Bend National Park, Texas, is shown above.

▲ An angular unconformity develops when new horizontal layers of sedimentary rock form on top of older sedimentary rock layers that have been folded by compression. An example of an angular unconformity at Siccar Point in southeastern Scotland is shown above.

▼ A disconformity develops when horizontal rock layers are exposed and eroded, and new horizontal layers of rock are deposited on the eroded surface. The disconformity shown below is in the Grand Canyon.

Matching Up Rock Layers

Suppose you're studying a layer of sandstone in Bryce Canyon in Utah. Later, when you visit Canyonlands National Park, Utah, you notice that a layer of sandstone there looks just like the sandstone in Bryce Canyon, 250 km away. Above the sandstone in the Canyonlands is a layer of limestone and then another sandstone layer. You return to Bryce Canyon and find the same sequence—sandstone, limestone, and sandstone. What do you infer? It's likely that you're looking at the same layers of rocks in two different locations. **Figure 15** shows that these rocks are parts of huge deposits that covered this whole area of the western United States. Geologists often can match up, or correlate, layers of rocks over great distances.

Evidence Used for Correlation It's not always easy to say that a rock layer exposed in one area is the same as a rock layer exposed in another area. Sometimes it's possible to walk along the layer for kilometers and prove that it's continuous. In other cases, such as at the Canyonlands area and Bryce Canyon as seen in **Figure 16,** the rock layers are exposed only where rivers have cut through overlying layers of rock and sediment. How can you show that the limestone sandwiched between the two layers of sandstone in Canyonlands is likely the same limestone as at Bryce Canyon? One way is to use fossil evidence. If the same types of fossils were found in the limestone layer in both places, it's a good indication that the limestone at each location is the same age, and, therefore, one continuous deposit.

✔ **Reading Check** *How do fossils help show that rocks at different locations belong to the same rock layer?*

Figure 15 These rock layers, exposed at Hopi Point in Grand Canyon National Park, Arizona, can be correlated, or matched up, with rocks from across large areas of the western United States.

Canyonlands National Park

Bryce Canyon National Park

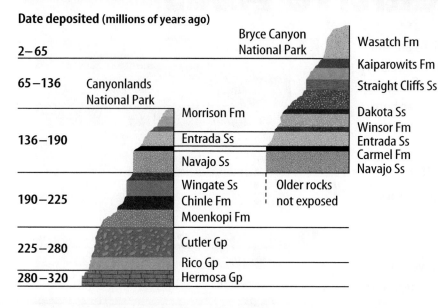

Date deposited (millions of years ago)

2–65	Bryce Canyon National Park — Wasatch Fm
	Kaiparowits Fm
65–136	Canyonlands National Park — Straight Cliffs Ss
	Morrison Fm — Dakota Ss
	Winsor Fm
136–190	Entrada Ss — Entrada Ss
	Carmel Fm
	Navajo Ss — Navajo Ss
	Wingate Ss — Older rocks
190–225	Chinle Fm — not exposed
	Moenkopi Fm
225–280	Cutler Gp
	Rico Gp
280–320	Hermosa Gp

Figure 16 Geologists have named the many rock layers, or formations, in Canyonlands and in Bryce Canyon, Utah. They also have correlated some formations between the two canyons.
List the labeled layers present at both canyons.

Can layers of rock be correlated in other ways? Sometimes determining relative ages isn't enough and other dating methods must be used. In Section 3, you'll see how the numerical ages of rocks can be determined and how geologists have used this information to estimate the age of Earth.

section 2 review

Summary

Superposition

● Superposition states that in undisturbed rock, the oldest layers are on the bottom.

Relative Ages

● Rock layers can be ranked by relative age.

Unconformities

● Angular unconformities are new layers deposited over tilted and eroded rock layers.

● Disconformities are gaps in the rock record.

● Nonconformities divide uplifted igneous or metamorphic rock from new sedimentary rock.

Matching Up Rock Layers

● Rocks from different areas may be correlated if they are part of the same layer.

Self Check

1. **Discuss** how to find the oldest paper in a stack of papers.

2. **Explain** the concept of relative age.

3. **Illustrate** a disconformity.

4. **Describe** one way to correlate similar rock layers.

5. **Think Critically** Explain the relationship between the concept of relative age and the principle of superposition.

Applying Skills

6. **Interpret data** to determine the oldest rock bed. A sandstone contains a 400-million-year-old fossil. A shale has fossils that are over 500 million years old. A limestone, below the sandstone, contains fossils between 400 million and 500 million years old. Which rock bed is oldest? Explain.

Relative Ages

Which of your two friends is older? To answer this question, you'd need to know their relative ages. You wouldn't need to know the exact age of either of your friends—just who was born first. The same is sometimes true for rock layers.

◖ Real-World Question

Can you determine the relative ages of rock layers?

Goals
■ **Interpret** illustrations of rock layers and other geological structures and determine the relative order of events.

Materials
paper   pencil

◖ Procedure

1. **Analyze Figures A** and **B.**
2. Make a sketch of **Figure A.** On it, identify the relative age of each rock layer, igneous intrusion, fault, and unconformity. For example, the shale layer is the oldest, so mark it with a 1. Mark the next-oldest feature with a 2, and so on.
3. Repeat step 2 for **Figure B.**

◖ Conclude and Apply

Figure A

1. **Identify** the type of unconformity shown. Is it possible that there were originally more layers of rock than are shown?
2. **Describe** how the rocks above the fault moved in relation to rocks below the fault.
3. **Hypothesize** how the hill on the left side of the figure formed.

Granite	Limestone
Sandstone	Shale

Figure B

4. Is it possible to conclude if the igneous intrusion on the left is older or younger than the unconformity nearest the surface?
5. **Describe** the relative ages of the two igneous intrusions. How did you know?
6. **Hypothesize** which two layers of rock might have been much thicker in the past.

Compare your results with other students' results. **For more help, refer to the** Science Skill Handbook.

Absolute Ages of Rocks

Absolute Ages

As you sort through your stack of magazines looking for that article about the car you saw, you decide that you need to restack them into a neat pile. By now, they're in a jumble and no longer in order of their relative age, as shown in **Figure 17.** How can you stack them so the oldest are on the bottom and the newest are on top? Fortunately, magazine dates are printed on the cover. Thus, stacking magazines in order is a simple process. Unfortunately, rocks don't have their ages stamped on them. Or do they? **Absolute age** is the age, in years, of a rock or other object. Geologists determine absolute ages by using properties of the atoms that make up materials.

Radioactive Decay

INTEGRATE Physics Atoms consist of a dense central region called the nucleus, which is surrounded by a cloud of negatively charged particles called electrons. The nucleus is made up of protons, which have a positive charge, and neutrons, which have no electric charge. The number of protons determines the identity of the element, and the number of neutrons determines the form of the element, or isotope. For example, every atom with a single proton is a hydrogen atom. Hydrogen atoms can have no neutrons, a single neutron, or two neutrons. This means that there are three isotopes of hydrogen.

✓ **Reading Check** *What particles make up an atom's nucleus?*

Some isotopes are unstable and break down into other isotopes and particles. Sometimes a lot of energy is given off during this process. The process of breaking down is called **radioactive decay.** In the case of hydrogen, atoms with one proton and two neutrons are unstable and tend to break down. Many other elements have stable and unstable isotopes.

What **You'll Learn**
- **Identify** how absolute age differs from relative age.
- **Describe** how the half-lives of isotopes are used to determine a rock's age.

Why **It's Important**

Events in Earth's history can be better understood if their absolute ages are known.

🔎 **Review Vocabulary**
isotopes: atoms of the same element that have different numbers of neutrons

New Vocabulary
- absolute age
- radioactive decay
- half-life
- radiometric dating
- uniformitarianism

Figure 17 The magazines that have been shuffled through no longer illustrate the principle of superposition.

Modeling Carbon-14 Dating

Procedure

1. Count out 80 **red jelly beans.**
2. Remove half the red jelly beans and replace them with **green jelly beans.**
3. Continue replacing half the red jelly beans with green jelly beans until only 5 red jelly beans remain. Count the number of times you replace half the red jelly beans.

Analysis

1. How did this activity model the decay of carbon-14 atoms?
2. How many half lives of carbon-14 did you model during this activity?
3. If the atoms in a bone experienced the same number of half lives as your jelly beans, how old would the bone be?

Figure 18 In beta decay, a neutron changes into a proton by giving off an electron. This electron has a lot of energy and is called a beta particle.

In the process of alpha decay, an unstable parent isotope nucleus gives off an alpha particle and changes into a new daughter product. Alpha particles contain two neutrons and two protons.

Alpha and Beta Decay In some isotopes, a neutron breaks down into a proton and an electron. This type of radioactive decay is called beta decay because the electron leaves the atom as a beta particle. The nucleus loses a neutron but gains a proton. When the number of protons in an atom is changed, a new element forms. Other isotopes give off two protons and two neutrons in the form of an alpha particle. Alpha and beta decay are shown in **Figure 18.**

Half-Life In radioactive decay reactions, the parent isotope undergoes radioactive decay. The daughter product is produced by radioactive decay. Each radioactive parent isotope decays to its daughter product at a certain rate. Based on this decay rate, it takes a certain period of time for one half of the parent isotope to decay to its daughter product. The **half-life** of an isotope is the time it takes for half of the atoms in the isotope to decay. For example, the half-life of carbon-14 is 5,730 years. So it will take 5,730 years for half of the carbon-14 atoms in an object to change into nitrogen-14 atoms. You might guess that in another 5,730 years, all of the remaining carbon-14 atoms will decay to nitrogen-14. However, this is not the case. Only half of the atoms of carbon-14 remaining after the first 5,730 years will decay during the second 5,730 years. So, after two half-lives, one fourth of the original carbon-14 atoms still remain. Half of them will decay during another 5,730 years. After three half-lives, one eighth of the original carbon-14 atoms still remain. After many half-lives, such a small amount of the parent isotope remains that it might not be measurable.

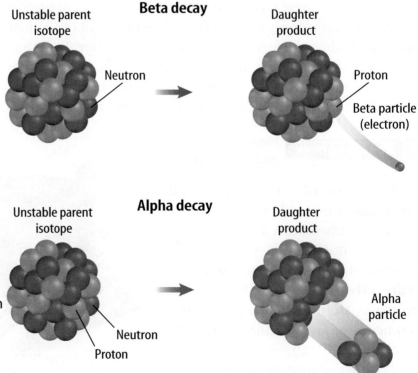

Beta decay

Unstable parent isotope — Neutron → Daughter product — Proton, Beta particle (electron)

Alpha decay

Unstable parent isotope — Neutron, Proton → Daughter product — Alpha particle

Radiometric Ages

Decay of radioactive isotopes is like a clock keeping track of time that has passed since rocks have formed. As time passes, the amount of parent isotope in a rock decreases as the amount of daughter product increases, as in **Figure 19.** By measuring the ratio of parent isotope to daughter product in a mineral and by knowing the half-life of the parent, in many cases you can calculate the absolute age of a rock. This process is called **radiometric dating.**

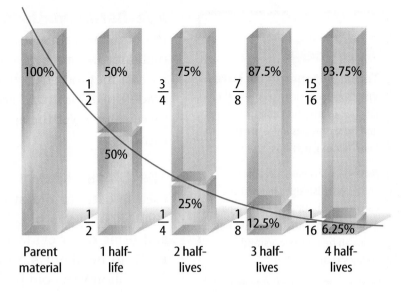

Figure 19 During each half-life, one half of the parent material decays to the daughter product. **Explain** *how one uses both parent and daughter material to estimate age.*

A scientist must decide which parent isotope to use when measuring the age of a rock. If the object to be dated seems old, then the geologist will use an isotope with a long half-life. The half-life for the decay of potassium-40 to argon-40 is 1.25 billion years. As a result, this isotope can be used to date rocks that are many millions of years old. To avoid error, conditions must be met for the ratios to give a correct indication of age. For example, the rock being studied must still retain all of the argon-40 that was produced by the decay of potassium-40. Also, it cannot contain any contamination of daughter product from other sources. Potassium-argon dating is good for rocks containing potassium, but what about other things?

Radiocarbon Dating Carbon-14 is useful for dating bones, wood, and charcoal up to 75,000 years old. Living things take in carbon from the environment to build their bodies. Most of that carbon is carbon-12, but some is carbon-14, and the ratio of these two isotopes in the environment is always the same. After the organism dies, the carbon-14 slowly decays. By determining the amounts of the isotopes in a sample, scientists can evaluate how much the isotope ratio in the sample differs from that in the environment. For example, during much of human history, people built campfires. The wood from these fires often is preserved as charcoal. Scientists can determine the amount of carbon-14 remaining in a sample of charcoal by measuring the amount of radiation emitted by the carbon-14 isotope in labs like the one in **Figure 20.** Once they know the amount of carbon-14 in a charcoal sample, scientists can determine the age of the wood used to make the fire.

Figure 20 Radiometric ages are determined in labs like this one.

Age Determinations Aside from carbon-14 dating, rocks that can be radiometrically dated are mostly igneous and metamorphic rocks. Most sedimentary rocks cannot be dated by this method. This is because many sedimentary rocks are made up of particles eroded from older rocks. Dating these pieces only gives the age of the preexisting rock from which it came.

The Oldest Known Rocks Radiometric dating has been used to date the oldest rocks on Earth. These rocks are about 3.96 billion years old. By determining the age of meteorites, and using other evidence, scientists have estimated the age of Earth to be about 4.5 billion years. Earth rocks greater than 3.96 billion years old probably were eroded or changed by heat and pressure.

✓ **Reading Check** *Why can't most sedimentary rocks be dated radiometrically?*

Applying Science

When did the Iceman die?

Carbon-14 dating has been used to date charcoal, wood, bones, mummies from Egypt and Peru, the Dead Sea Scrolls, and the Italian Iceman. The Iceman was found in 1991 in the Italian Alps, near the Austrian border. Based on carbon-14 analysis, scientists determined that the Iceman is 5,300 years old. Determine approximately in what year the Iceman died.

Half-Life of Carbon-14	
Percent Carbon-14	Years Passed
100	0
50	5,730
25	11,460
12.5	17,190
6.25	22,920
3.125	

Reconstruction of Iceman

Identifying the Problem

The half-life chart shows the decay of carbon-14 over time. Half-life is the time it takes for half of a sample to decay. Fill in the years passed when only 3.125 percent of carbon-14 remain. Is there a point at which no carbon-14 would be present? Explain.

Solving the Problem

1. Estimate, using the data table, how much carbon-14 still was present in the Iceman's body that allowed scientists to determine his age.
2. If you had an artifact that originally contained 10.0 g of carbon-14, how many grams would remain after 17,190 years?

Uniformitarianism

Can you imagine trying to determine the age of Earth without some of the information you know today? Before the discovery of radiometric dating, many people estimated that Earth is only a few thousand years old. But in the 1700s, Scottish scientist James Hutton estimated that Earth is much older. He used the principle of **uniformitarianism.** This principle states that Earth processes occurring today are similar to those that occurred in the past. Hutton's principle is often paraphrased as "the present is the key to the past."

Figure 21 The rugged highlands of Scotland were shaped by erosion and uplift.

Hutton observed that the processes that changed the landscape around him were slow, and he inferred that they were just as slow throughout Earth's history. Hutton hypothesized that it took much longer than a few thousand years to form the layers of rock around him and to erode mountains that once stood kilometers high. **Figure 21** shows Hutton's native Scotland, a region shaped by millions of years of geologic processes.

Today, scientists recognize that Earth has been shaped by two types of change: slow, everyday processes that take place over millions of years, and violent, unusual events such as the collision of a comet or asteroid about 65 million years ago that might have caused the extinction of the dinosaurs.

section 3 review

Summary

Absolute Ages
- The absolute age is the actual age of an object.

Radioactive Decay
- Some isotopes are unstable and decay into other isotopes and particles.
- Decay is measured in half-lives, the time it takes for half of a given isotope to decay.

Radiometric Ages
- By measuring the ratio of parent isotope to daughter product, one can determine the absolute age of a rock.
- Living organisms less than 75,000 years old can be dated using carbon-14.

Uniformitarianism
- Processes observable today are the same as the processes that took place in the past.

Self Check

1. **Evaluate** the age of rocks. You find three undisturbed rock layers. The middle layer is 120 million years old. What can you say about the ages of the layers above and below it?
2. **Determine** the age of a fossil if it had only one eighth of its original carbon-14 content remaining.
3. **Explain** the concept of uniformitarianism.
4. **Describe** how radioactive isotopes decay.
5. **Think Critically** Why can't scientists use carbon-14 to determine the age of an igneous rock?

Applying Math

6. **Make and use a table** that shows the amount of parent material of a radioactive element that is left after four half-lives if the original parent material had a mass of 100 g.

Model and Invent

Trace Fossils

Goals
■ **Construct** a model of trace fossils.
■ **Describe** the information that you can learn from looking at your model.

Possible Materials
construction paper
wire
plastic (a fairly rigid type)
scissors
plaster of paris
toothpicks
sturdy cardboard
clay
pipe cleaners
glue

Safety Precautions

▶ Real-World Question

Trace fossils can tell you a lot about the activities of organisms that left them. They can tell you how an organism fed or what kind of home it had. How can you model trace fossils that can provide information about the behavior of organisms? What materials can you use to model trace fossils? What types of behavior could you show with your trace fossil model?

▶ Make a Model

1. **Decide** how you are going to make your model. What materials will you need?

2. **Decide** what types of activities you will demonstrate with your model. Were the organisms feeding? Resting? Traveling? Were they predators? Prey? How will your model indicate the activities you chose?

3. What is the setting of your model? Are you modeling the organism's home? Feeding areas? Is your model on land or water? How can the setting affect the way you build your model?

4. Will you only show trace fossils from a single species or multiple species? If you include more than one species, how will you provide evidence of any interaction between the species?

Check the Model Plans

1. Compare your plans with those of others in your class. Did other groups mention details that you had forgotten to think about? Are there any changes you would like to make to your plan before you continue?

2. Make sure your teacher approves your plan before you continue.

◉ Test Your Model

1. Following your plan, construct your model of trace fossils.
2. Have you included evidence of all the behaviors you intended to model?

◉ Analyze Your Data

1. **Evaluate** Now that your model is complete, do you think that it adequately shows the behaviors you planned to demonstrate? Is there anything that you think you might want to do differently if you were going to make the model again?

2. **Describe** how using different kinds of materials might have affected your model. Can you think of other materials that would have allowed you to show more detail than you did?

◉ Conclude and Apply

1. **Compare and contrast** your model of trace fossils with trace fossils left by real organisms. Is one more easily interpreted than the other? Explain.

2. **List** behaviors that might not leave any trace fossils. Explain.

*C*ommunicating Your Data

Ask other students in your class or another class to look at your model and describe what information they can learn from the trace fossils. Did their interpretations agree with what you intended to show?

The World's Oldest Fish Story

A catch-of-the-day set science on its ears

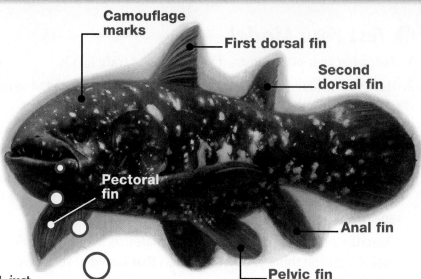

Camouflage marks

First dorsal fin

Second dorsal fin

Pectoral fin

Anal fin

Pelvic fin

Some scientists call the coelacanth "Old Four Legs." It got its nickname because the fish has paired fins that look something like legs.

O n a December day in 1938, just before Christmas, Marjorie Courtenay-Latimer went to say hello to her friends on board a fishing boat that had just returned to port in South Africa. Courtenay-Latimer, who worked at a museum, often went aboard her friends' ship to check out the catch. On this visit, she received a surprise Christmas present—an odd-looking fish. As soon as the woman spotted its strange blue fins among the piles of sharks and rays, she knew it was special.

Courtenay-Latimer took the fish back to her museum to study it. "It was the most beautiful fish I had ever seen, five feet long, and a pale mauve blue with iridescent silver markings," she later wrote. Courtenay-Latimer sketched it and sent the drawing to a friend of hers, J. L. B. Smith.

Smith was a chemistry teacher who was passionate about fish. After a time, he realized it was a coelacanth (SEE luh kanth). Fish experts knew that coelacanths had first appeared on Earth 400 million years ago. But the experts thought the fish were extinct. People had found fossils of coelacanths, but

no one had seen one alive. It was assumed that the last coelacanth species had died out 65 million years ago. They were wrong. The ship's crew had caught one by accident.

Smith figured there might be more living coelacanths. So he decided to offer a reward for anyone who could find a living specimen. After 14 years of silence, a report came in that a coelacanth had been caught off the east coast of Africa.

Today, scientists know that there are at least several hundred coelacanths living in the Indian Ocean, just east of central Africa. Many of these fish live near the Comoros Islands. The coelacanths live in underwater caves during the day but move out at night to feed. The rare fish are now a protected species. With any luck, they will survive for another hundred million years.

Write a short essay describing the discovery of the coelacanths and describe the reaction of scientists to this discovery.

Science Online

For more information, visit bookg.msscience.com/oops

Reviewing Main Ideas

Section 1 **Fossils**

1. Fossils are more likely to form if hard parts of the dead organisms are buried quickly.

2. Some fossils form when original materials that made up the organisms are replaced with minerals. Other fossils form when remains are subjected to heat and pressure, leaving only a carbon film behind. Some fossils are the tracks or traces left by ancient organisms.

Section 2 **Relative Ages of Rocks**

1. The principle of superposition states that, in undisturbed layers, older rocks lie underneath younger rocks.

2. Unconformities, or gaps in the rock record, are due to erosion or periods of time during which no deposition occurred.

3. Rock layers can be correlated using rock types and fossils.

Section 3 **Absolute Ages of Rocks**

1. Absolute dating provides an age in years for the rocks.

2. The half-life of a radioactive isotope is the time it takes for half of the atoms of the isotope to decay into another isotope.

Visualizing Main Ideas

Copy and complete the following concept map on fossils.

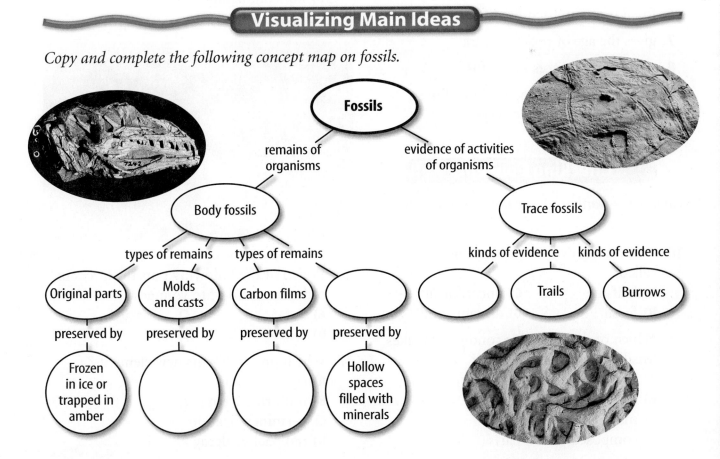

Using Vocabulary

absolute age p. 139
carbon film p. 126
cast p. 127
fossil p. 125
half-life p. 140
index fossil p. 129
mold p. 127
permineralized remains p. 126

principle of superposition p. 132
radioactive decay p. 139
radiometric dating p. 141
relative age p. 133
unconformity p. 134
uniformitarianism p. 143

Write an original sentence using the vocabulary word to which each phrase refers.

1. thin film of carbon preserved as a fossil

2. older rocks lie under younger rocks

3. processes occur today as they did in the past

4. gap in the rock record

5. time needed for half the atoms to decay

6. fossil organism that lived for a short time

7. gives the age of rocks in years

8. minerals fill spaces inside fossil

9. a copy of a fossil produced by filling a mold with sediment or crystals

Checking Concepts

Choose the word or phrase that best answers the question.

10. What is any evidence of ancient life called?
    A) half-life
    B) fossil
    C) unconformity
    D) disconformity

11. Which of the following conditions makes fossil formation more likely?
    A) buried slowly
    B) attacked by scavengers
    C) made of hard parts
    D) composed of soft parts

12. What are cavities left in rocks when a shell or bone dissolves called?
    A) casts
    B) molds
    C) original remains
    D) carbon films

13. To say "the present is the key to the past" is a way to describe which of the following principles?
    A) superposition
    B) succession
    C) radioactivity
    D) uniformitarianism

14. A fault can be useful in determining which of the following for a group of rocks?
    A) absolute age
    B) index age
    C) radiometric age
    D) relative age

15. Which of the following is an unconformity between parallel rock layers?
    A) angular unconformity
    B) fault
    C) disconformity
    D) nonconformity

Use the illustration below to answer question 16.

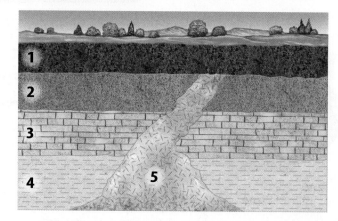

16. Which of the following puts the layers in order from oldest to youngest?
    A) 5-4-3-2-1
    B) 1-2-3-4-5
    C) 2-3-4-5-1
    D) 4-3-2-5-1

17. Which process forms new elements?
    A) superposition
    B) uniformitarianism
    C) permineralization
    D) radioactive decay

Science Online bookg.msscience.com/vocabulary_puzzlemaker

Thinking Critically

18. Explain why the fossil record of life on Earth is incomplete. Give some reasons why.

19. Infer Suppose a lava flow was found between two sedimentary rock layers. How could you use the lava flow to learn about the ages of the sedimentary rock layers? *(Hint: Most lava contains radioactive isotopes.)*

20. Infer Suppose you're correlating rock layers in the western United States. You find a layer of volcanic ash deposits. How can this layer help you in your correlation over a large area?

21. Recognize Cause and Effect Explain how some woolly mammoths could have been preserved intact in frozen ground. What conditions must have persisted since the deaths of these animals?

22. Classify each of the following fossils in the correct category in the table below: *dinosaur footprint, worm burrow, dinosaur skull, insect in amber, fossil woodpecker hole,* and *fish tooth.*

Types of Fossils	
Trace Fossils	**Body Fossils**
Do not write in this book.	

23. Compare and contrast the three different kinds of unconformities. Draw sketches of each that illustrate the features that identify them.

24. Describe how relative and absolute ages differ. How might both be used to establish ages in a series of rock layers?

25. Discuss uniformitarianism in the following scenario. You find a shell on the beach, and a friend remembers seeing a similar fossil while hiking in the mountains. What does this suggest about the past environment of the mountain?

Performance Activities

26. Illustrate Create a model that allows you to explain how to establish the relative ages of rock layers.

27. Use a Classification System Start your own fossil collection. Label each find as to type, approximate age, and the place where it was found. Most state geological surveys can provide you with reference materials on local fossils.

Applying Math

28. Calculate how many half-lives have passed in a rock containing one-eighth the original radioactive material and seven-eighths of the daughter product.

Use the graphs below to answer question 29.

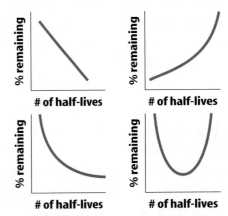

29. Interpret Data Which of the above curves best illustrates radioactive decay?

Part 1 | Multiple Choice

Record your answers on the answer sheet provided by your teacher or on a sheet of paper.

Use the photo below to answer question 1.

1. Which type of fossil preservation is shown above?
   A. trace fossil
   B. original remains
   C. carbon film
   D. permineralized remains

2. Which principle states that the oldest rock layer is found at the bottom in an undisturbed stack of rock layers?
   A. half-life
   B. absolute dating
   C. superposition
   D. uniformitarianism

3. Which type of scientist studies fossils?
   A. meteorologist
   B. chemist
   C. astronomer
   D. paleontologist

4. Which are the remains of species that existed on Earth for relatively short periods of time, were abundant, and were widespread geographically?
   A. trace fossils
   B. index fossils
   C. carbon films
   D. body fossils

5. Which term means matching up rock layers in different places?
   A. superposition
   B. correlation
   C. uniformitarianism
   D. absolute dating

6. Which of the following is least likely to be found as a fossil?
   A. clam shell
   B. shark tooth
   C. snail shell
   D. jellyfish imprint

7. Which type of fossil preservation is a thin carbon silhouette of the original organism?
   A. cast
   B. carbon film
   C. mold
   D. permineralized remains

8. Which isotope is useful for dating wood and charcoal that is less than about 75,000 years old?
   A. carbon-14
   B. potassium-40
   C. uranium-238
   D. argon-40

Use the diagram below to answer questions 9–11.

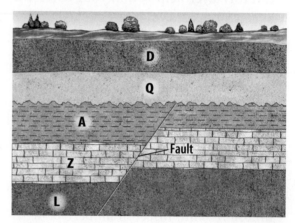

9. Which sequence of letters describes the rock layers in the diagram from oldest to youngest?
   A. D, Q, A, Z, L
   B. L, Z, A, Q, D
   C. Z, L, A, D, Q
   D. Q, D, L, Z, A

10. What does the wavy line between layers A and Q represent?
    A. a disconformity
    B. a fault
    C. a nonconformity
    D. an angular unconformity

11. Which of the following correctly describes the relative age of the fault?
    A. younger than A, but older than Q
    B. younger than Z, but older than L
    C. younger than Q, but older than A
    D. younger than D, but older than Q

Part 2 | Short Response/Grid In

Record your answers on the answer sheet provided by your teacher or on a sheet of paper.

12. What is a fossil?

13. How is a fossil cast different from a fossil mold?

14. Describe the principle of uniformitarianism.

15. Explain how the original remains of an insect can be preserved as a fossil in amber.

16. Why do scientists hypothesize that Earth is about 4.5 billion years old?

17. Describe the process of radioactive decay. Use the terms *isotope, nucleus,* and *half-life* in your answer.

Use the table below to answer questions 18–20.

Number of Half-lives	Parent Isotope Remaining (%)
1	100
2	X
3	25
4	12.5
5	Y

18. What value should replace the letter X in the table above?

19. What value should replace the letter Y in the table above?

20. Explain the relationship between the number of half-lives that have elapsed and the amount of parent isotope remaining.

21. Compare and contrast the three types of unconformities.

22. Why are index fossils useful for estimating the age of rock layers?

Part 3 | Open Ended

Record your answers on a sheet of paper.

23. Why are fossils important? What information do they provide?

24. List three different types of trace fossils. Explain how each type forms.

Examine the graph below and answer questions 25–27.

Relationship Between Sediment Burial Rate and Potential for Remains to Become Fossils

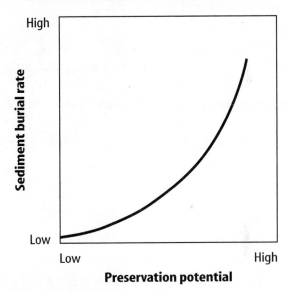

25. How does the potential for remains to be preserved change as the rate of burial by sediment increases?

26. Why do you think this relationship exists?

27. What other factors affect the potential for the remains of organisms to become fossils?

28. How could a fossil of an organism that lived in ocean water millions of years ago be found in the middle of North America?

Test-Taking Tip

Check It Again Double check your answers before turning in the test.

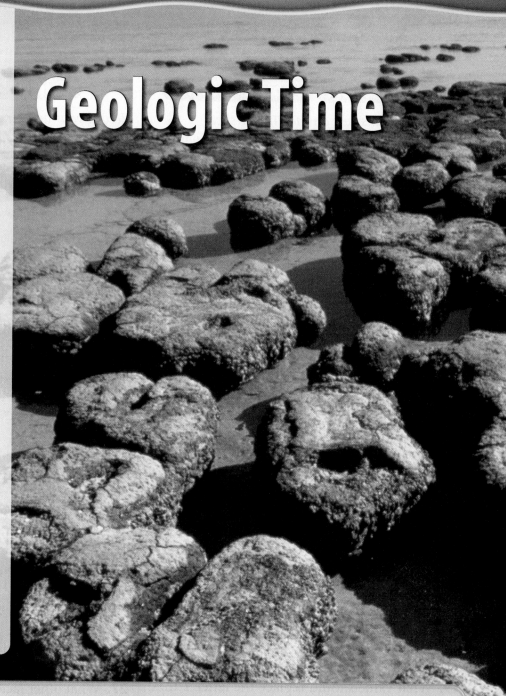

Geologic Time

The BIG Idea

Scientists use units of geologic time to interpret the history of life on Earth.

SECTION 1
Life and Geologic Time

Main Idea Fossils provide evidence that life on Earth has evolved throughout geologic time.

SECTION 2
Early Earth History

Main Idea Primitive life forms existed on Earth during Precambrian time and the Paleozoic Era.

SECTION 3
Middle and Recent Earth History

Main Idea Life forms continued to evolve through the Mesozoic Era and the current Cenozoic Era.

Looking at the Past

The stromatolites in the picture hardly have changed since they first appeared 3.5 billion years ago. Looking at these organisms today allows us to imagine what the early Earth might have looked like. In this chapter, you will see how much Earth has changed over time, even as some parts remain the same.

Science Journal Describe how an animal or plant might change if Earth becomes hotter in the next million years.

Start-Up Activities

Survival Through Time

Environments include the living and nonliving things that surround and affect organisms. Whether or not an organism survives in its environment depends upon its characteristics. Only if an organism survives until adulthood can it reproduce and pass on its characteristics to its offspring. In this lab, you will use a model to find out how one characteristic can determine whether individuals can survive in an environment.

1. Cut 15 pieces each of green, orange, and blue yarn into 3-cm lengths.

2. Scatter them on a sheet of green construction paper.

3. Have your partner use a pair of tweezers to pick up as many pieces as possible in 15 s.

4. **Think Critically** In your Science Journal, discuss which colors your partner selected. Which color was least selected? Suppose that the construction paper represents grass, the yarn pieces represent insects, and the tweezers represent an insect-eating bird. Which color of insect do you predict would survive to adulthood?

Geological Time Make the following Foldable to help you identify the major events in each era of geologic time.

STEP 1 Fold the top of a vertical piece of paper down and the bottom up to divide the paper into thirds.

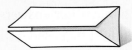

STEP 2 Turn the paper horizontally; **unfold and label** the three columns as shown.

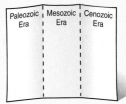

Read for Main Ideas As you read the chapter, list at least three major events that occurred in each era. Keep the events in chronological order. For each event, note the period in which it took place.

Preview this chapter's content and activities at
bookg.msscience.com

Get Ready to Read

Questions and Answers

1 Learn It! Knowing how to find answers to questions will help you on reviews and tests. Some answers can be found in the textbook, while other answers require you to go beyond the textbook. These answers might be based on knowledge you already have or things you have experienced.

2 Practice It! Read the excerpt below. Answer the following questions and then discuss them with a partner.

> The Paleozoic Era, or era of ancient life, began about 544 million years ago and ended about 248 million years ago. Traces of life are much easier to find in Paleozoic rocks than in Precambrian rocks.
>
> —*from page 164*

- What does the term *Paleozoic* mean?
- What does the term *Precambrian* mean?
- Why are traces of life easier to find in Paleozoic rocks than in Precambrian rocks?

3 Apply It! Look at some questions in the chapter review. Which questions can be answered directly from the text? Which require you to go beyond the text?

Target Your Reading

Use this to focus on the main ideas as you read the chapter.

1 **Before you read** the chapter, respond to the statements below on your worksheet or on a numbered sheet of paper.

- Write an **A** if you **agree** with the statement.
- Write a **D** if you **disagree** with the statement.

2 **After you read** the chapter, look back to this page to see if you've changed your mind about any of the statements.

- If any of your answers changed, explain why.
- Change any false statements into true statements.
- Use your revised statements as a study guide.

Sciencenline

Print out a worksheet of this page at bookg.msscience.com

Before You Read A or D		Statement	After You Read A or D
	1	The fossil record shows that species have changed over geologic time.	
	2	Geologic time units are based on life-forms that lived only during certain periods of time.	
	3	Eras are longer than eons.	
	4	Precambrian time is the shortest part of Earth's history.	
	5	No life-forms existed on Earth during Precambrian time.	
	6	Oxygen gas has always been a major component of Earth's atmosphere throughout geologic time.	
	7	All dinosaurs were large, slow-moving, cold-blooded reptiles.	
	8	Dinosaurs lived during the Mesozoic Era.	
	9	Many scientists hypothesize that a comet or asteroid collision with Earth, ended Mesozoic Era.	

Life and Geologic Time

as you read

What **You'll Learn**

- **Explain** how geologic time can be divided into units.
- **Relate** changes of Earth's organisms to divisions on the geologic time scale.
- **Describe** how plate tectonics affects species.

Why **It's Important**

The life and landscape around you are the products of change through geologic time.

Review Vocabulary

fossils: remains, traces, or imprints of prehistoric organisms

New Vocabulary

- geologic time scale
- eon
- era
- period
- epoch
- organic evolution
- species
- natural selection
- trilobite
- Pangaea

Geologic Time

A group of students is searching for fossils. By looking in rocks that are hundreds of millions of years old, they hope to find many examples of trilobites (TRI loh bites) so that they can help piece together a puzzle. That puzzle is to find out what caused the extinction of these organisms. **Figure 1** shows some examples of what they are finding. The fossils are small, and their bodies are divided into segments. Some of them seem to have eyes. Could these interesting fossils be trilobites?

Trilobites are small, hard-shelled organisms that crawled on the seafloor and sometimes swam through the water. Most ranged in size from 2 cm to 7 cm in length and from 1 cm to 3 cm in width. They are considered to be index fossils because they lived over vast regions of the world during specific periods of geologic time.

The Geologic Time Scale The appearance or disappearance of types of organisms throughout Earth's history marks important occurrences in geologic time. Paleontologists have been able to divide Earth's history into time units based on the life-forms that lived only during certain periods. This division of Earth's history makes up the **geologic time scale.** However, sometimes fossils are not present, so certain divisions of the geologic time scale are based on other criteria.

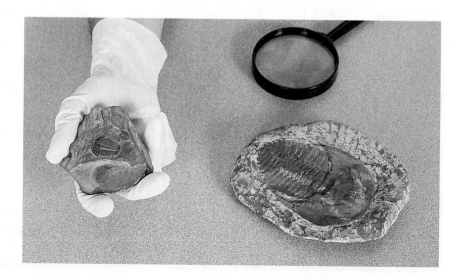

Figure 1 Many sedimentary rocks in the United States are rich in invertebrate fossils such as these trilobites.

Major Subdivisions of Geologic Time The oldest rocks on Earth contain no fossils. Then, for many millions of years after the first appearance of fossils, the fossil record remained sparse. Later in Earth's history came an explosion in the abundance and diversity of organisms. These organisms left a rich fossil record. As shown in **Figure 2,** four major subdivisions of geologic time are used—eons, eras, periods, and epochs. The longest subdivisions—**eons**—are based upon the abundance of certain fossils.

Reading Check *What are the major subdivisions of geologic time?*

Next to eons, the longest subdivisions are the **eras,** which are marked by major, striking, and worldwide changes in the types of fossils present. For example, at the end of the Mesozoic Era, many kinds of invertebrates, birds, mammals, and reptiles became extinct.

Eras are subdivided into periods. **Periods** are units of geologic time characterized by the types of life existing worldwide at the time. Periods can be divided into smaller units of time called **epochs.** Epochs also are characterized by differences in life-forms, but some of these differences can vary from continent to continent. Epochs of periods in the Cenozoic Era have been given specific names. Epochs of other periods usually are referred to simply as early, middle, or late. Epochs are further subdivided into units of shorter duration.

Dividing Geologic Time There is a limit to how finely geologic time can be subdivided. It depends upon the kind of rock record that is being studied. Sometimes it is possible to distinguish layers of rock that formed during a single year or season. In other cases, thick stacks of rock that have no fossils provide little information that could help in subdividing geologic time.

Figure 2 Scientists have divided the geologic time scale into subunits based upon the appearance and disappearance of types of organisms.
Explain *how the even blocks in this chart can be misleading.*

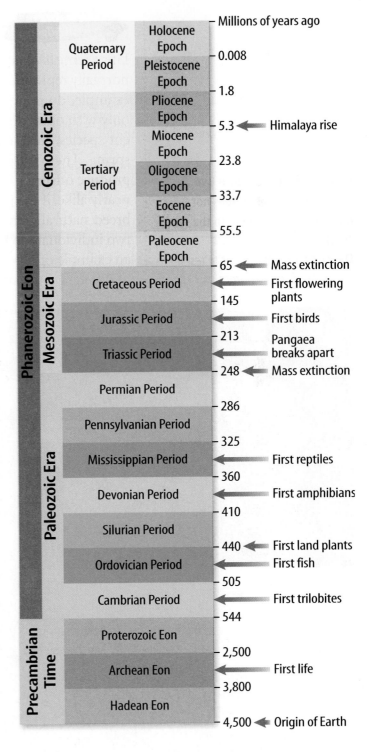

Organic Evolution

The fossil record shows that species have changed over geologic time. This change through time is known as **organic evolution.** According to most theories about organic evolution, environmental changes can affect an organism's survival. Those organisms that are not adapted to changes are less likely to survive or reproduce. Over time, the elimination of individuals that are not adapted can cause changes to species of organisms.

INTEGRATE Life Science

Species Many ways of defining the term species (SPEE sheez) have been proposed. Life scientists often define a **species** as a group of organisms that normally reproduces only with other members of their group. For example, dogs are a species because dogs mate and reproduce only with other dogs. In some rare cases, members of two different species, such as lions and tigers, can mate and produce offspring. These offspring, however, are usually sterile and cannot produce offspring of their own. Even though two organisms look nearly alike, if the populations they each come from do not interbreed naturally and produce offspring that can reproduce, the two individuals do not belong to the same species. **Figure 3** shows an example of two species that look similar to each other but live in different areas and do not mate naturally with each other.

Figure 3 Just because two organisms look alike does not mean that they belong to the same species.
Describe *an experiment to test if these lizards are separate species.*

The coast horned lizard lives along the coast of central and southern California.

The desert horned lizard lives in arid regions of the southwestern United States.

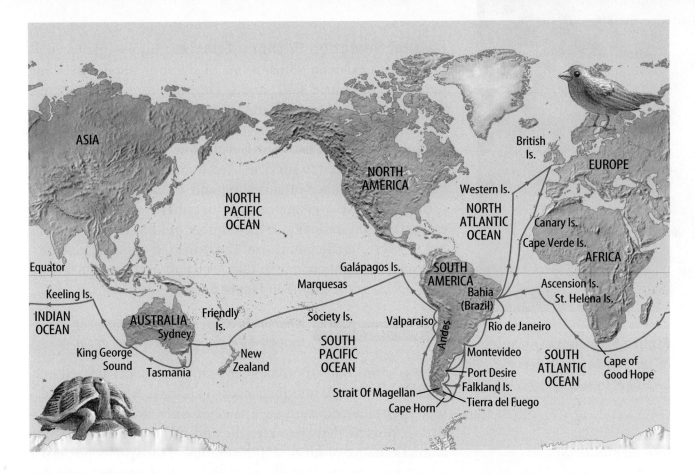

Figure 4 Charles Darwin sailed around the world between 1831 and 1836 aboard the HMS *Beagle* as a naturalist. On his journey he saw an abundance of evidence for natural selection, especially on the Galápagos Islands off the western coast of South America.

Natural Selection Charles Darwin was a naturalist who sailed around the world from 1831 to 1836 to study biology and geology. **Figure 4** shows a map of his journey. With some of the information about the plants and animals he observed on this trip in mind, he later published a book about the theory of evolution by natural selection.

In his book, he proposed that **natural selection** is a process by which organisms with characteristics that are suited to a certain environment have a better chance of surviving and reproducing than organisms that do not have these characteristics. Darwin knew that many organisms are capable of producing more offspring than can survive. This means that organisms compete with each other for resources necessary for life, such as food and living space. He also knew that individual organisms within the same species could be different, or show variations, and that these differences could help or hurt the individual organism's chance of surviving.

Some organisms that were well suited to their environment lived longer and had a better chance of producing offspring. Organisms that were poorly adapted to their environment produced few or no offspring. Because many characteristics are inherited, the characteristics of organisms that are better adapted to the environment get passed on to offspring more often. According to Darwin, this can cause a species to change over time.

Figure 5 Giraffes can eat leaves off the branches of tall trees because of their long necks.

Natural Selection Within a Species Suppose that an animal species exists in which a few of the individuals have long necks, but most have short necks. The main food for the animal is the leafy foliage on trees in the area. What happens if the climate changes and the area becomes dry? The lower branches of the trees might not have any leaves. Now which of the animals will be better suited to survive? Clearly, the long-necked animals have a better chance of surviving and reproducing. Their offspring will have a greater chance of inheriting the important characteristic. Gradually, as the number of long-necked animals becomes greater, the number of short-necked animals decreases. The species might change so that nearly all of its members have long necks, as the giraffe in **Figure 5** has.

Reading Check *What might happen to the population of animals if the climate became wet again?*

It is important to notice that individual, short-necked animals didn't change into long-necked animals. A new characteristic becomes common in a species only if some members already possess that characteristic and if the trait increases the animal's chance of survival. If no animal in the species possessed a long neck in the first place, a long-necked species could not have evolved by means of natural selection.

Artificial Selection Humans have long used the principle of artificial selection when breeding domestic animals. By carefully choosing individuals with desired characteristics, animal breeders have created many breeds of cats, dogs, cattle, and chickens. **Figure 6** shows the great variety of cats produced by artificial selection.

Figure 6 Cat breeders have succeeded in producing a great variety of cats by using the principle of artificial selection.

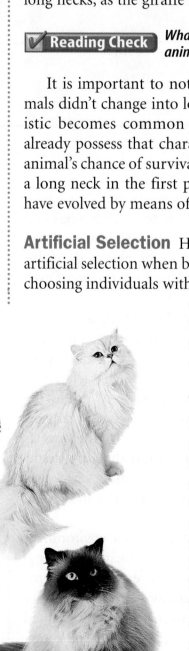

The Evolution of New Species
Natural selection explains how characteristics change and how new species arise. For example, if the short-necked animals migrated to a different location, they might have survived. They could have continued to reproduce in the new location, eventually developing enough different characteristics from the long-necked animals that they might not be able to breed with each other. At this point, at least one new species would have evolved.

Trilobites

Remember the trilobites? The term *trilobite* comes from the structure of the hard outer skeleton or exoskeleton. The exoskeleton of a **trilobite** consists of three lobes that run the length of the body. As shown in **Figure 7,** the trilobite's body also has a head (cephalon), a segmented middle section (thorax), and a tail (pygidium).

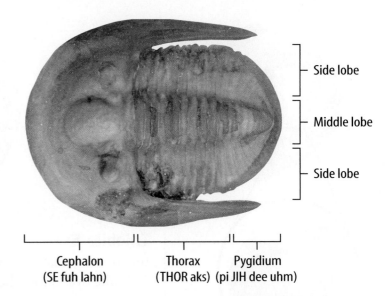

Cephalon
(SE fuh lahn)

Thorax
(THOR aks)

Pygidium
(pi JIH dee uhm)

Figure 7 The trilobite's body was divided into three lobes that run the length of the body—two side lobes and one middle lobe.

Changing Characteristics of Trilobites Trilobites inhabited Earth's oceans for more than 200 million years. Throughout the Paleozoic Era, some species of trilobites became extinct and other new species evolved. Species of trilobites that lived during one period of the Paleozoic Era showed different characteristics than species from other periods of this era. As **Figure 8** shows, paleontologists can use these different characteristics to demonstrate changes in trilobites through geologic time. These changes can tell you about how different trilobites from different periods lived and responded to changes in their environments.

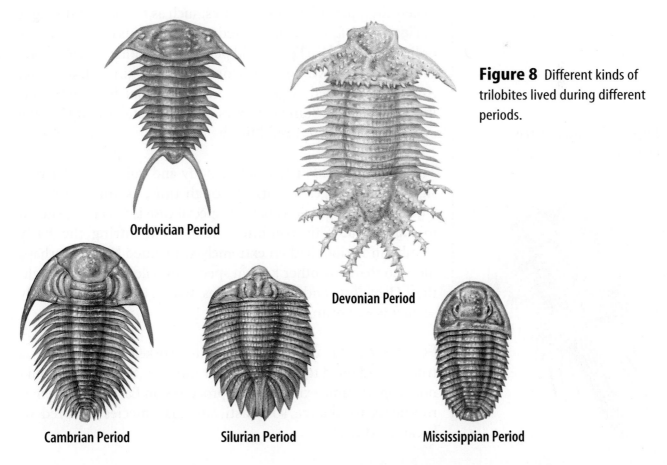

Ordovician Period

Devonian Period

Cambrian Period

Silurian Period

Mississippian Period

Figure 8 Different kinds of trilobites lived during different periods.

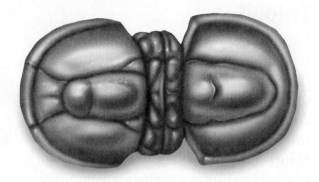

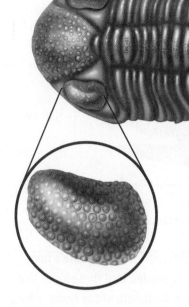

Figure 9 Trilobites had many different types of eyes. Some had eyes that contained hundreds of small circular lenses, somewhat like an insect. The blind trilobite (right) had no eyes.

Figure 10 *Olenellus* is one of the most primitive trilobite species.

Trilobite Eyes Trilobites, shown in **Figure 9,** might have been the first organisms that could view the world with complex eyes. Trilobite eyes show the result of natural selection. The position of the eyes on an organism gives clues about where it must have lived. Eyes that are located toward the front of the head indicate an organism that was adapted for active swimming. If the eyes are located toward the back of the head, the organism could have been a bottom dweller. In most species of trilobites, the eyes were located midway on the head—a compromise for an organism that was adapted for crawling on the seafloor and swimming in the water.

Over time, the eyes in trilobites changed. In many trilobite species, the eyes became progressively smaller until they completely disappeared. Blind trilobites, such as the one on the right in **Figure 9,** might have burrowed into sediments on the seafloor or lived deeper than light could penetrate. In other species, however, the eyes became more complex. One kind of trilobite, *Aeglina*, developed large compound eyes that had numerous individual lenses. Some trilobites developed stalks that held the eyes upward. Where would this be useful?

Trilobite Bodies The trilobite body and tail also underwent significant changes in form through time, as you can see in **Figure 8** on the previous page. A special case is *Olenellus*, shown in **Figure 10.** This trilobite, which lived during the Early Cambrian Period, had an extremely segmented body—perhaps more so than any other known species of trilobite. It is thought that *Olenellus*, and other species that have so many body segments, are primitive trilobites.

Fossils Show Changes Trilobite exoskeletons changed as trilobites adapted to changing environments. Species that could not adapt became extinct. What processes on Earth caused environments to change so drastically that species adapted or became extinct?

Plate Tectonics and Earth History

Plate tectonics is one possible answer to the riddle of trilobite extinction. Earth's moving plates caused continents to collide and separate many times. Continental collisions formed mountains and closed seas caught between continents. Continental separations created wider, deeper seas between continents. By the end of the Paleozoic Era, sea levels had dropped and the continents had come together to form one giant landmass, the supercontinent **Pangaea** (pan JEE uh). Because trilobites lived in the oceans, their environment was changed or destroyed. **Figure 11** shows the arrangement of continents at the end of the Paleozoic Era. What effect might these changes have had on the trilobite populations?

Not all scientists accept the above explanation for the extinctions at the end of the Paleozoic Era, and other possibilities—such as climate change—have been proposed. As in all scientific debates, you must consider the evidence carefully and come to conclusions based on the evidence.

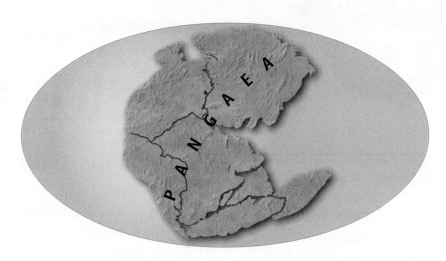

Figure 11 The amount of shallow water environment was reduced when Pangaea formed. **Describe** *how this change affected organisms that lived along the coasts of continents.*

section 1 review

Summary

Geologic Time

- Earth's history is divided into eons, eras, periods, and epochs, based on fossils.

Organic Evolution

- The fossil record indicates that species have changed over time.

- Charles Darwin proposed natural selection to explain change in species.

- In natural selection, organisms best suited to their environments survive and produce the most offspring.

Trilobites

- Trilobites were abundant in the Paleozoic fossil record and can be used as index fossils.

Plate Tectonics and Earth History

- Continents moving through time have influenced the environments of past organisms.

Self Check

1. **Discuss** how fossils relate to the geologic time scale.
2. **Infer** how plate tectonics might lead to extinction.
3. **Infer** how the eyes of a trilobite show how it lived.
4. **Explain** how paleontologists use trilobite fossils as index fossils for various geologic time periods.
5. **Think Critically** Aside from moving plates, what other factors could cause an organism's environment to change? How would this affect species?

Applying Skills

6. **Recognize Cause and Effect** Answer the questions below.

   a. How does natural selection cause evolutionary change to take place?

   b. How could the evolution of a characteristic within one species affect the evolution of a characteristic within another species? Give an example.

Early Earth History

as you read

What You'll Learn

- **Identify** characteristic Precambrian and Paleozoic life-forms.
- **Draw** conclusions about how species adapted to changing environments in Precambrian time and the Paleozoic Era.
- **Describe** changes in Earth and its life-forms at the end of the Paleozoic Era.

Why It's Important

The Precambrian includes most of Earth's history.

🔍 Review Vocabulary

life: state of being in which one grows, reproduces, and maintains a constant internal environment

New Vocabulary

- Precambrian time
- cyanobacteria
- Paleozoic Era

Precambrian Time

It may seem strange, but **Figure 12** is probably an accurate picture of Earth's first billion years. Over the next 3 billion years, simple life-forms began to colonize the oceans.

Look again at the geologic time scale shown in **Figure 2. Precambrian** (pree KAM bree un) **time** is the longest part of Earth's history and includes the Hadean, Archean, and Proterozoic Eons. Precambrian time lasted from about 4.5 billion years ago to about 544 million years ago. The oldest rocks that have been found on Earth are about 4 billion years old. However, rocks older than about 3.5 billion years are rare. This probably is due to remelting and erosion.

Although the Precambrian was the longest interval of geologic time, relatively little is known about the organisms that lived during this time. One reason is that many Precambrian rocks have been so deeply buried that they have been changed by heat and pressure. Many fossils can't withstand these conditions. In addition, most Precambrian organisms didn't have hard parts that otherwise would have increased their chances to be preserved as fossils.

Figure 12 During the early Precambrian, Earth was a lifeless planet with many volcanoes.

Lava flow

Lava flow

Ash

Ash deposits

Ocean

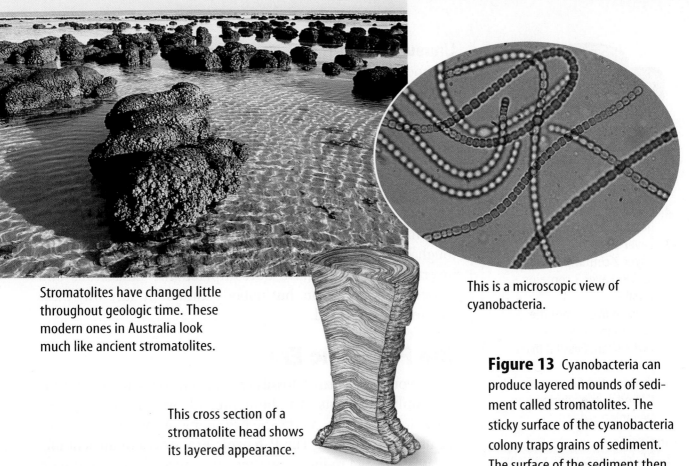

Stromatolites have changed little throughout geologic time. These modern ones in Australia look much like ancient stromatolites.

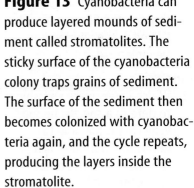

This cross section of a stromatolite head shows its layered appearance.

This is a microscopic view of cyanobacteria.

Figure 13 Cyanobacteria can produce layered mounds of sediment called stromatolites. The sticky surface of the cyanobacteria colony traps grains of sediment. The surface of the sediment then becomes colonized with cyanobacteria again, and the cycle repeats, producing the layers inside the stromatolite.

Early Life Many studies of the early history of life involve ancient stromatolites (stroh MA tuh lites). **Figure 13** shows stromatolites, which are layered mats formed by cyanobacteria colonies. **Cyanobacteria** are blue-green algae thought to be one of the earliest forms of life on Earth. Cyanobacteria first appeared about 3.5 billion years ago. They contained chlorophyll and used photosynthesis. This is important because during photosynthesis, they produced oxygen, which helped change Earth's atmosphere. Following the appearance of cyanobacteria, oxygen became a major atmospheric gas. Also of importance was that the ozone layer in the atmosphere began to develop, shielding Earth from ultraviolet rays. It is hypothesized that these changes allowed species of single-celled organisms to evolve into more complex organisms.

Reading Check *What atmospheric gas is produced by photosynthesis?*

Animals without backbones, called invertebrates (ihn VUR tuh brayts), appeared toward the end of Precambrian time. Imprints of invertebrates have been found in late Precambrian rocks, but because these early invertebrates were soft bodied, they weren't often preserved as fossils. Because of this, many Precambrian fossils are trace fossils.

INTEGRATE Chemistry

Earth's First Air
Cyanobacteria are thought to have been one of the mechanisms by which Earth's early atmosphere became richer in oxygen. Research the composition of Earth's early atmosphere and where these gases probably came from. Record your findings in your Science Journal.

Mini LAB

Dating Rock Layers with Fossils

Procedure

1. Draw three rock layers.
2. Number the layers 1 to 3, bottom to top.
3. Layer 1 contains fossil A. Layer 2 contains fossils A and B. Layer 3 contains fossil C.
4. Fossil A lived from the Cambrian through the Ordovician. Fossil B lived from the Ordovician through the Silurian. Fossil C lived in the Silurian and Devonian.

Analysis

1. Which layers were you able to date to a specific period?
2. Why isn't it possible to determine during which specific period the other layers formed?

Figure 14 This giant predatory fish lived in seas that were present in North America during the Devonian Period. It grew to about 6 m in length.

Unusual Life-Forms A group of animals with shapes similar to modern jellyfish, worms, and soft corals was living late in Precambrian time. Fossils of these organisms were first found in the Ediacara Hills in southern Australia. This group of organisms has become known as the Ediacaran (ee dee uh KAR un) fauna. **Figure 15** shows some of these fossils.

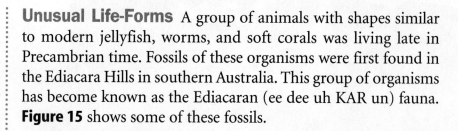

Reading Check *What modern organisms do some Ediacaran organisms resemble?*

Ediacaran animals were bottom dwellers and might have had tough outer coverings like air mattresses. Trilobites and other invertebrates might have outcompeted the Ediacarans and caused their extinction, but nobody knows for sure why these creatures disappeared.

The Paleozoic Era

As you have learned, fossils are unlikely to form if organisms have only soft parts. An abundance of organisms with hard parts, such as shells, marks the beginning of the Paleozoic (pay lee uh ZOH ihk) Era. The **Paleozoic Era,** or era of ancient life, began about 544 million years ago and ended about 248 million years ago. Traces of life are much easier to find in Paleozoic rocks than in Precambrian rocks.

Paleozoic Life Because warm, shallow seas covered large parts of the continents during much of the Paleozoic Era, many of the life-forms scientists know about were marine, meaning they lived in the ocean. Trilobites were common, especially early in the Paleozoic. Other organisms developed shells that were easily preserved as fossils. Therefore, the fossil record of this era contains abundant shells. However, invertebrates were not the only animals to live in the shallow, Paleozoic seas.

Vertebrates, or animals with backbones, also evolved during this era. The first vertebrates were fishlike creatures without jaws. Armoured fish with jaws such as the one shown in **Figure 14** lived during the Devonian Period. Some of these fish were so huge that they could eat large sharks with their powerful jaws. By the Devonian Period, forests had appeared and vertebrates began to adapt to land environments, as well.

Figure 15

A variety of 600-million-year-old fossils—known as Ediacaran (eed ee uh KAR un) fauna—have been found on every continent except Antarctica. These unusual organisms were originally thought to be descendants of early animals such as jellyfish, worms, and coral. Today, paleontologists debate whether these organisms were part of the animal kingdom or belonged to an entirely new kingdom whose members became extinct about 545 million years ago.

DICKENSONIA (dihk un suh NEE uh) Impressions of *Dickensonia,* a bottom-dwelling wormlike creature, have been discovered. Some are nearly one meter long.

RANGEA (rayn JEE uh) As it lay rooted in sea-bottom sediments, *Rangea* may have snagged tiny bits of food by filtering water through its body.

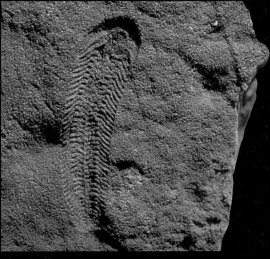

SPRIGGINA (sprih GIHN uh) Some scientists hypothesize that the four-centimeter-long *Spriggina* was a type of crawling, segmented organism. Others suggest that it sat upright while attached to the seafloor.

CYCLOMEDUSA (si kloh muh DEW suh) Although it looks a lot like a jellyfish, *Cyclomedusa* may have had more in common with modern sea anemones. Some paleontologists, however, hypothesize that it is unrelated to any living organism.

Figure 16 Amphibians probably evolved from fish like *Panderichthys* (pan dur IHK theez), which had leglike fins and lungs.

Science Online

Topic: Paleozoic Life

Visit bookg.msscience.com for Web links to information about Paleozoic life.

Activity Prepare a presentation on the organisms of one period of the Paleozoic Era. Describe a few animals from different groups, including how and where they lived. Are any of these creatures alive today, and if not, when did they become extinct?

Life on Land Based on their structure, paleontologists know that many ancient fish had lungs as well as gills. Lungs enabled these fish to live in water with low oxygen levels—when needed they could swim to the surface and breathe air. Today's lungfish also can get oxygen from the water through gills and from the air through lungs.

One kind of ancient fish had lungs and leglike fins, which were used to swim and crawl around on the ocean bottom. Paleontologists hypothesize that amphibians might have evolved from this kind of fish, shown in **Figure 16.** The characteristics that helped animals survive in oxygen-poor waters also made living on land possible. Today, amphibians live in a variety of habitats in water and on land. They all have at least one thing in common, though. They must lay their eggs in water or moist places.

Reading Check *What are some characteristics of the fish from which amphibians might have evolved?*

By the Pennsylvanian Period, some amphibians evolved an egg with a membrane that protected it from drying out. Because of this, these animals, called reptiles, no longer needed to lay eggs in water. Reptiles also have skin with hard scales that prevent loss of body fluids. This adaptation enables them to survive farther from water and in relatively dry climates, as shown in **Figure 17,** where many amphibians cannot live.

Figure 17 Reptiles have scaly skins that allow them to live in dry places.

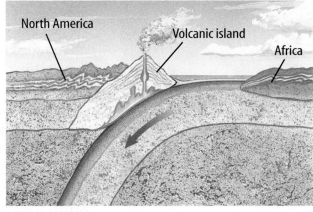

More than 375 million years ago, volcanic island chains formed in the ocean and were pushed against the coast as Africa moved toward North America.

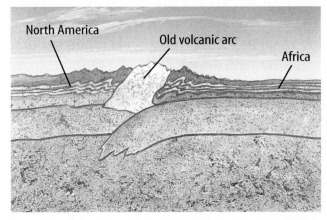

About 375 million years ago, the African plate collided with the North American plate, forming mountains on both continents.

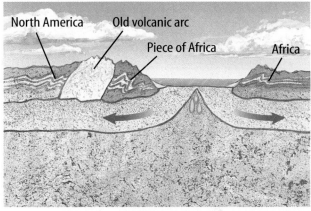

About 200 million years ago, the Atlantic Ocean opened up, separating the two continents.

Mountain Building Several mountain-building episodes occurred during the Paleozoic Era. The Appalachian Mountains, for example, formed during this time. This happened in several stages, as shown in **Figure 18.** The first mountain-building episode occurred as the ocean separating North America from Europe and Africa closed. Several volcanic island chains that had formed in the ocean collided with the North American Plate, as shown in the top picture of **Figure 18.** The collision of the island chains generated high mountains.

The next mountain-building episode was a result of the African Plate colliding with the North American Plate, as shown in the left picture of **Figure 18.** When Africa and North America collided, rock layers were folded and faulted. Some rocks originally deposited near the eastern coast of the North American Plate were pushed along faults as much as 65 km westward by the collision. Sediments were uplifted to form an immense mountain belt, part of which still remains today.

Figure 18 The Appalachian Mountains formed in several stages.
Infer *how these movements affected species in the Appalachians.*

Figure 19 Hyoliths were organisms that became extinct at the end of the Paleozoic Era.

End of an Era At the end of the Paleozoic Era, more than 90 percent of all marine species and 70 percent of all land species died off. **Figure 19** shows one such animal. The cause of these extinctions might have been changes in climate and a lowering of sea level.

Near the end of the Permian Period, the continental plates came together and formed the supercontinent Pangaea. Glaciers formed over most of its southern part. The slow, gradual collision of continental plates caused mountain building. Mountain-building processes caused seas to close and deserts to spread over North America and Europe. Many species, especially marine organisms, couldn't adapt to these changes, and became extinct.

Other Hypotheses Other explanations also have been proposed for this mass extinction. During the late Paleozoic Era, volcanoes were extremely active. If the volcanic activity was great enough, it could have affected the entire globe. Another recent theory is similar to the one proposed to explain the extinction of dinosaurs. Perhaps a large asteroid or comet collided with Earth some 248 million years ago. This event could have caused widespread extinctions just as many paleontologists suggest happened at the end of the Mesozoic Era, 65 million years ago. Perhaps the extinction at the end of the Paleozoic Era was caused by several or all of these events happening at about the same time.

section 2 review

Summary

Precambrian Time

- Precambrian time covers almost 4 billion years of Earth history, but little is known about the organisms of this time.
- Cyanobacteria were among the earliest life-forms.

The Paleozoic Era

- Invertebrates developed shells and other hard parts, leaving a rich fossil record.
- Vertebrates—animals with backbones—appeared during this era.
- Plants and amphibians first moved to land during the Paleozoic Era.
- Adaptations in reptiles allow them to move away from water for reproduction.
- Geologic events at the end of the Paleozoic Era led to a mass extinction.

Self Check

1. **List** the geologic events that ended the Paleozoic Era.
2. **Infer** how geologic events at the end of the Paleozoic Era might have caused extinctions.
3. **Discuss** the advance that allowed reptiles to reproduce away from water. Why was this an advantage?
4. **Identify** the major change in life-forms that occurred at the end of Precambrian time.
5. **Think Critically** How did cyanobacteria aid the evolution of complex life on land? Do you think cyanobacteria are as significant to this process today as they were during Precambrian time?

Applying Skills

6. **Use a Database** Research trilobites and describe these organisms and their habitats in your Science Journal. Include hand-drawn illustrations and compare them with the illustrations in your references.

Changing Species

In this lab, you will observe how adaptation within a species might cause the evolution of a particular trait, leading to the development of a new species.

◉ Real-World Question

How might adaptation within a species cause the evolution of a particular trait?

Goals
- ■ **Model** adaptation within a species.

Materials
deck of playing cards

◉ Procedure

1. **Remove** all of the kings, queens, jacks, and aces from a deck of playing cards.
2. Each remaining card represents an individual in a population of animals called "varimals." The number on each card represents the height of the individual.
3. **Calculate** the average height of the population of varimals represented by your cards.
4. Suppose varimals eat grass, shrubs, and leaves from trees. A drought causes many of these plants to die. All that's left are a few tall trees. Only varimals at least 6 units tall can reach the leaves on these trees.
5. All the varimals under 6 units leave the area or die from starvation. Discard all of the cards with a number less than 6. Calculate the new average height of the varimals.
6. **Shuffle** the deck of remaining cards.
7. **Draw** two cards at a time. Each pair represents a pair of varimals that will mate.

8. The offspring of each pair reaches the average height of its parents. Calculate and record the height of each offspring.
9. Discard all parents and offspring under 8 units tall and repeat steps 6–8. Now calculate the new average height of varimals. Include both the parents and offspring in your calculation.

◉ Conclude and Apply

1. **Describe** how the height of the population changed.
2. **Explain** If you hadn't discarded the shortest varimals, would the average height of the population have changed as much?
3. Suppose the offspring grew to the height of one of its parents. How would the results change in each of the following scenarios?
   a. The height value for the offspring is chosen by coin toss.
   b. The height value for the offspring is whichever parent is tallest.
4. **Explain** If there had been no variation in height before the droughts occurred, would the species have been able to evolve?

Middle and Recent Earth History

as you read

What You'll Learn

■ **Compare and contrast** characteristic life-forms in the Mesozoic and Cenozoic Eras.
■ **Explain** how changes caused by plate tectonics affected organisms during the Mesozoic Era.
■ **Identify** when humans first appeared on Earth.

Why It's Important

Many important groups of animals, like birds and mammals, appeared during the Mesozoic Era.

🔍 **Review Vocabulary**
dinosaur: a reptile from one of two orders that dominated the Mesozoic Era

New Vocabulary
● Mesozoic Era
● Cenozoic Era

The Mesozoic Era

Dinosaurs have captured people's imaginations since their bones first were unearthed more than 150 years ago. Dinosaurs and other interesting animals lived during the Mesozoic Era, which was between 248 and 65 million years ago. The Mesozoic Era also was marked by rapid movement of Earth's plates.

The Breakup of Pangaea The **Mesozoic** (meh zuh ZOH ihk) **Era,** or era of middle life, was a time of many changes on Earth. At the beginning of the Mesozoic Era, all continents were joined as a single landmass called Pangaea, as shown in **Figure 11.**

Pangaea separated into two large landmasses during the Triassic Period, as shown in **Figure 20.** The northern mass was Laurasia (law RAY zhuh), and Gondwanaland (gahn DWAH nuh land) was the southern landmass. As the Mesozoic Era continued, Laurasia and Gondwanaland broke apart and eventually formed the present-day continents.

Species that had adapted to the new environments survived the mass extinction at the end of the Paleozoic Era. Recall that a reptile's skin helps it retain bodily fluids. This characteristic, along with their shelled eggs, enabled reptiles to adapt readily to the drier climate of the Mesozoic Era. Reptiles became the most conspicuous animals on land by the Triassic Period.

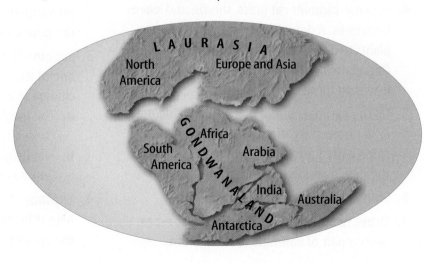

Figure 20 At the end of the Triassic Period, Pangaea began to break up into the northern supercontinent, Laurasia, and the southern supercontinent, Gondwanaland.

Dinosaurs What were the dinosaurs like? Dinosaurs ranged in height from less than 1 m to enormous creatures like *Apatosaurus* and *Tyrannosaurus*. The first small dinosaurs appeared during the Triassic Period. Larger species appeared during the Jurassic and Cretaceous Periods. Throughout the Mesozoic Era, new species of dinosaurs evolved and other species became extinct.

Dinosaurs Were Active Studying fossil footprints sometimes allows paleontologists to calculate how fast animals walked or ran. Some dinosaur tracks indicate that these animals were much faster runners than you might think. *Gallimimus* was 4 m long and could reach speeds of 65 km/h—as fast as a modern racehorse.

Some studies also indicate that dinosaurs might have been warm blooded, not cold blooded like present-day reptiles. The evidence that leads to this conclusion has to do with their bone structure. Slices through some cold-blooded animal bones show rings similar to growth rings in trees. The bones of some dinosaurs don't show this ring structure. Instead, they are similar to bones found in modern mammals, as you can see in **Figure 21.**

Reading Check *Why do some paleontologists think that dinosaurs were warm blooded?*

These observations indicate that some dinosaurs might have been warm-blooded, fast-moving animals somewhat like present-day mammals and birds. They might have been quite different from present-day reptiles.

Good Mother Dinosaurs The fossil record also indicates that some dinosaurs nurtured their young and traveled in herds in which the adults surrounded their young.

One such dinosaur is *Maiasaura*. This dinosaur built nests in which it laid its eggs and raised its offspring. Nests have been found in relatively close clusters, indicating that more than one family of dinosaurs built in the same area. Some fossils of hatchlings have been found near adult animals, leading paleontologists to think that some dinosaurs nurtured their young. In fact, *Maiasaura* hatchlings might have stayed in the nest while they grew in length from about 35 cm to more than 1 m.

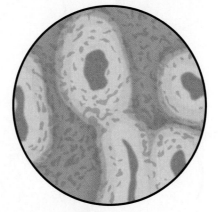
Dinosaur bone

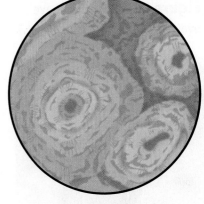
Mammal bone

Figure 21 Some dinosaur bones show structural features that are like mammal bones, leading some paleontologists to think that dinosaurs were warm blooded like mammals.

Science Online

Topic: Warm Versus Cold
Visit bookg.msscience.com for Web links to information about dinosaurs.

Activity Work with a partner to research the debate on warm-blooded versus cold-blooded dinosaurs. Present your finding to the class in the form of a debate. Be sure to cover the main points of disagreement between the two sides.

Figure 22 Birds might have evolved from dinosaurs.

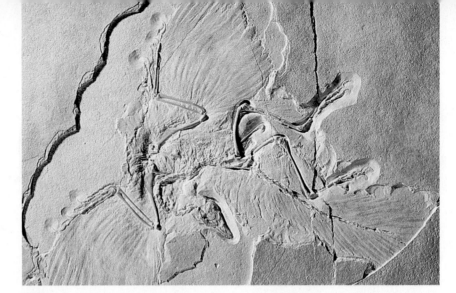

B Considered one of the world's most priceless fossils, *Archaeopteryx*, above, was first found in a limestone quarry in Germany in 1861.

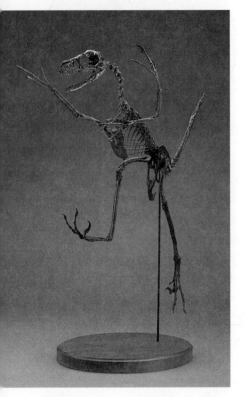

A *Bambiraptor feinberger*, above, is a 75-million-year-old member of a family of meat-eating dinosaurs thought by some paleontologists to be closely related to birds.

Figure 23 The earliest mammals were small creatures that resembled today's mice and shrews.

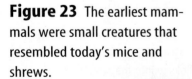

Birds Birds appeared during the Jurassic Period. Some paleontologists think that birds evolved from small, meat-eating dinosaurs much like *Bambiraptor feinberger* in **Figure 22A.** The earliest bird, *Archaeopteryx,* shown in **Figure 22B,** had wings and feathers. However, because *Archaeopteryx* had features not shared with modern birds, scientists know it was not a direct ancestor of today's birds.

Mammals Mammals first appeared in the Triassic Period. The earliest mammals were small, mouselike creatures, as shown in **Figure 23.** Mammals are warm-blooded vertebrates that have hair covering their bodies. The females produce milk to feed their young. These two characteristics have enabled mammals to survive in many changing environments.

Gymnosperms During most of the Mesozoic Era, gymnosperms (JIHM nuh spurmz), which first appeared in the Paleozoic Era, dominated the land. Gymnosperms are plants that produce seeds but not flowers. Many gymnosperms are still around today. These include pines and ginkgo trees.

Angiosperms Angiosperms (AN jee uh spurmz), or flowering plants, first evolved during the Cretaceous Period. Angiosperms produce seeds with hard outer coverings.

Because their seeds are enclosed and protected, angiosperms can live in many environments. Angiosperms are the most diverse and abundant land plants today. Present-day angiosperms that evolved during the Mesozoic Era include magnolia and oak trees.

End of an Era The Mesozoic Era ended about 65 million years ago with a major extinction of land and marine species. Many groups of animals, including the dinosaurs, disappeared suddenly at this time. Many paleontologists hypothesize that a comet or asteroid collided with Earth, causing a huge cloud of dust and smoke to rise into the atmosphere, blocking out the Sun. Without sunlight the plants died, and all the animals that depended on these plants also died. Not everything died, however. All the organisms that you see around you today are descendants of the survivors of the great extinction at the end of the Mesozoic Era.

Applying Math Calculate Percentages

CALCULATING EXTINCTION BY USING PERCENTAGES At the end of the Cretaceous Period, large numbers of plants and animals became extinct. Scientists still are trying to understand why some types of plants and animals survived while others died out. Looking at data about amphibians, reptiles, and mammals that lived during the Cretaceous Period, can you determine what percentage of amphibians survived this mass extinction?

Solution

1 *This is what you know:*

Animal Extinctions

Animal Type	Groups Living Before Extinction Event (n)	Groups Left After Extinction Event (t)
Amphibians	12	4
Reptiles	63	30
Mammals	24	8

2 *This is what you need to find out:*

p = the percentage of amphibian groups that survived the Cretaceous extinction

3 *This is the equation you need to use:*

- $p = t / n \times 100$
- Both t and n are shown on the above chart.

4 *Substitute the known values:*

$p = 4 / 12 \times 100 = 33.3\%$

Practice Problems

1. Using the same equation as demonstrated above, calculate the percentage of reptiles and then the percentage of mammals that survived. Which type of animal was least affected by the extinction?

2. What percentage of all groups survived?

For more practice, visit bookg.msscience.com/math_practice

The Cenozoic Era

The **Cenozoic** (se nuh ZOH ihk) **Era,** or era of recent life, began about 65 million years ago and continues today. Many mountain ranges in North and South America and Europe began to form in the Cenozoic Era. In the late Cenozoic, the climate became much cooler and ice ages occurred. The Cenozoic Era is subdivided into two periods. The first of these is the Tertiary Period. The present-day period is the Quaternary Period. It began about 1.8 million years ago.

Reading Check *What happened to the climate during the late Cenozoic Era?*

Times of Mountain Building Many mountain ranges formed during the Cenozoic Era. These include the Alps in Europe and the Andes in South America. The Himalaya, shown in **Figure 24,** formed as India moved northward and collided with Asia. The collision crumpled and thickened Earth's crust, raising the highest mountains presently on Earth. Many people think the growth of these mountains has helped create cooler climates worldwide.

Figure 24 The Himalaya extend along the India-Tibet border and contain some of the world's tallest mountains. India drifted north and finally collided with Asia, forming the Himalaya.

Further Evolution of Mammals

Throughout much of the Cenozoic Era, expanding grasslands favored grazing plant eaters like horses, camels, deer, and some elephants. Many kinds of mammals became larger. Horses evolved from small, multi-toed animals into the large, hoofed animals of today. However, not all mammals remained on land. Ancestors of the present-day whales and dolphins evolved to live in the sea.

As Australia and South America separated from Antarctica during the continuing breakup of the continents, many species became isolated. They evolved separately from life-forms in other parts of the world. Evidence of this can be seen today in Australia's marsupials. Marsupials are mammals such as kangaroos, koalas, and wombats (shown in **Figure 25**) that carry their young in a pouch.

Your species, *Homo sapiens*, probably appeared about 140,000 years ago. Some people suggest that the appearance of humans could have led to the extinction of many other mammals. As their numbers grew, humans competed for food that other animals relied upon. Also, fossil bones and other evidence indicate that early humans were hunters.

Figure 25 The wombat is one of many Australian marsupials. As a result of human activities, the number and range of wombats have diminished.

section 3 review

Summary

The Mesozoic Era

- During the Triassic Period, Pangaea split into two continents.
- Dinosaurs were the dominant land animals of the Mesozoic Era.
- Birds, mammals, and flowering plants all appeared during this era.
- The Mesozoic Era ended 65 million years ago with a mass extinction.

The Cenozoic Era

- The Cenozoic Era has been a mountain-building period with cooler climates.
- Mammals became dominant with many new life-forms appearing after the dinosaurs disappeared.
- Humans also appeared in the Cenozoic Era, probably about 140,000 years ago.

Self Check

1. **List** the era, period, and epoch in which *Homo sapiens* first appeared.
2. **Discuss** whether mammals became more or less abundant after the extinction of the dinosaurs, and explain why.
3. **Infer** how seeds with a hard outer covering enabled angiosperms to survive in a wide variety of climates.
4. **Explain** why some paleontologists hypothesize that dinosaurs were warm-blooded animals.
5. **Think Critically** How could two species that evolved on separate continents have many similarities?

Applying Math

6. **Convert Units** A fossil mosasaur, a giant marine reptile, measured 9 m in length and had a skull that measured 45 cm in length. What fraction of the mosasaur's total length did the skull account for? Compare your length with the mosasaur's length.

Use the Internet

Disc🕐vering the Past

Goals

- **Gather** information about fossils found in your area.
- **Communicate** details about fossils found in your area.
- **Synthesize** information from sources about the fossil record and the changes in your area over time.

Data Source

Science🖱nline

Visit **bookg.msscience.com/ internet_lab** for more information about fossils and changes over geologic time and for data collected by other students.

▶ *Real-World Question*

Imagine what your state was like millions of years ago. What animals might have been roaming around the spot where you now sit? Can you picture a *Tyrannosaurus rex* roaming the area that is now your school? The animals and plants that once inhabited your region might have left some clues to their identity—fossils. Scientists use fossils to piece together what Earth looked like in the geologic past. Fossils can help determine whether an area used to be dry land or underwater. Fossils can help uncover clues about how plants and animals have evolved over the course of time. Using the resources of the Internet and by sharing data with your peers, you can start to discover how North America has changed through time. How has your area changed over geologic time? How might the area where you are now living have looked thousands or millions of years ago? Do you think that the types of animals and plants have changed much over time? Form a hypothesis concerning the change in organisms and geography from long ago to the present day in your area.

Fossils in Your Area					
Fossil Name	Plant or Animal Fossil	Age of Fossils	Details About Plant or Animal Fossil	Location of Fossil	Additional Information
		Do not write in this book.			

▶ *Make a Plan*

1. **Determine** the age of the rocks that make up your area. Were they formed during Precambrian time, the Paleozoic Era, the Mesozoic Era, or the Cenozoic Era?

2. Gather information about the plants and animals found in your area during one of the above geologic time intervals. Find specific information on when, where, and how the fossil organisms lived. If no fossils are known from your area, find out information about the fossils found nearest your area.

▶ *Follow Your Plan*

1. Make sure your teacher approves your plan before you start.

2. Go to bookg.msscience.com/internet_lab to post your data in the table. Add any additional information you think is important to understanding the fossils found in your area.

▶ *Analyze Your Data*

1. What present-day relatives of prehistoric animals or plants exist in your area?

2. How have the organisms in your area changed over time? Is your hypothesis supported? Why or why not?

3. What other information did you discover about your area's climate or environment from the geologic time period you investigated?

▶ *Conclude and Apply*

1. **Describe** the plant and animal fossils that have been discovered in your area. What clues did you discover about the environment in which these organisms lived? How do these compare to the environment of your area today?

2. **Infer** from the fossil organisms found in your area what the geography and climate were like during the geologic time period you chose.

Communicating Your Data

Find this lab using the link below.

Science Online

bookg.msscience.com/internet_lab

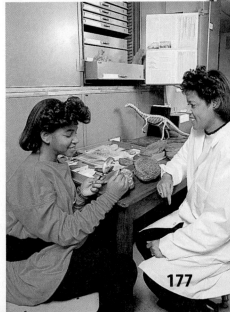

SCIENCE Stats

Extinct!

Did you know...

...The saber-toothed cat lived in the Americas from about 1.6 million to 8,000 years ago. *Smilodon*, the best-known saber-toothed cat, was among the most ferocious carnivores. It had large canine teeth, about 15 cm long, which it used to pierce the flesh of its prey.

Applying Math How many years did *Smilodon* live in the Americas before it became extinct?

Saber-toothed cat

Woolly mammoth

Great Mass Extinctions of Species

Periods	Percent extinction

...The woolly mammoth lived in the cold tundra regions during the Ice Age. It looked rather like an elephant with long hair, had a mass between 5,300 kg and 7,300 kg, and was between 3 m and 4 m tall.

Write About It

Visit bookg.msscience.com/science_stats **to research extinct animals. Trace the origins of each of the species and learn how long its kind existed on Earth.**

Reviewing Main Ideas

Section 1   Life and Geologic Time

1. Geologic time is divided into eons, eras, periods, and epochs.

2. Divisions within the geologic time scale are based largely on major evolutionary changes in organisms.

3. Plate movements affect organic evolution.

Section 2   Early Earth History

1. Cyanobacteria evolved during Precambrian time. Trilobites, fish, and corals were abundant during the Paleozoic Era.

2. Plants and animals began to move onto land during the middle of the Paleozoic Era.

3. The Paleozoic Era was a time of mountain building. The Appalachian Mountains formed when several islands and finally Africa collided with North America.

4. At the end of the Paleozoic Era, many marine invertebrates became extinct.

Section 3   Middle and Recent Earth History

1. Reptiles and gymnosperms were dominant land life-forms in the Mesozoic Era. Mammals and angiosperms began to dominate the land in the Cenozoic Era.

2. Pangaea broke apart during the Mesozoic Era. Many mountain ranges formed during the Cenozoic Era.

Visualizing Main Ideas

Copy and complete the concept map on geologic time using the following choices: Cenozoic, Trilobites in oceans, Mammals common, Paleozoic, Dinosaurs roam Earth, *and* Abundant gymnosperms.

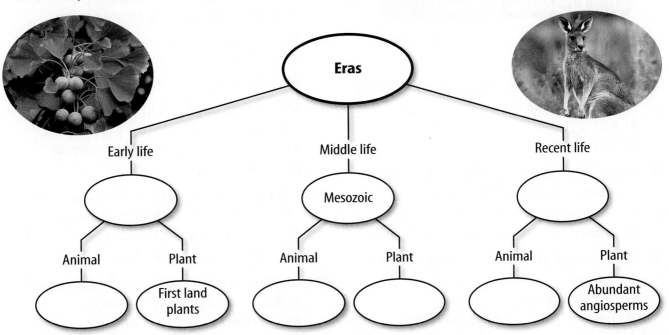

Using Vocabulary

Cenozoic Era p. 174
cyanobacteria p. 163
eon p. 155
epoch p. 155
era p. 155
geologic time scale p. 154
Mesozoic Era p. 170
natural selection p. 157

organic evolution p. 156
Paleozoic Era p. 164
Pangaea p. 161
period p. 155
Precambrian time p. 162
species p. 156
trilobite p. 159

Fill in the blank with the correct word or words.

1. A change in the hereditary features of a species over a long period is _____.

2. A record of events in Earth history is the _____.

3. The largest subdivision of geologic time is the _____.

4. The process by which the best-suited individuals survive in their environment is _____.

5. A group of individuals that normally breed only among themselves is a(n) _____.

Checking Concepts

Choose the word or phrase that best completes the sentence.

6. How many millions of years ago did the era in which you live begin?
   A) 650
   B) 245
   C) 1.6
   D) 65

7. What is the process by which better-suited organisms survive and reproduce?
   A) endangerment
   B) extinction
   C) gymnosperm
   D) natural selection

8. During what period did the most recent ice age occur?
   A) Pennsylvanian
   B) Triassic
   C) Tertiary
   D) Quaternary

9. What is the next smaller division of geologic time after the era?
   A) period
   B) stage
   C) epoch
   D) eon

10. What was one of the earliest forms of life?
    A) gymnosperm
    B) cyanobacterium
    C) angiosperm
    D) dinosaur

Use the illustration below to answer question 11.

11. Consider the undisturbed rock layers in the figure above. If fossil X were a *Tyrannosaurus rex* bone, and fossil Y were a trilobite; then fossil Z could be which of the following?
    A) stromatolite
    B) sabre-tooth cat
    C) angiosperm
    D) *Homo sapiens*

12. During which era did the dinosaurs live?
    A) Mesozoic
    B) Paleozoic
    C) Miocene
    D) Cenozoic

13. Which type of plant has seeds without protective coverings?
    A) angiosperms
    B) apples
    C) gymnosperms
    D) magnolias

14. Which group of plants evolved during the Mesozoic Era and is dominant today?
    A) gymnosperms
    B) angiosperms
    C) ginkgoes
    D) algae

15. In which era did the Ediacaran fauna live?
    A) Precambrian
    B) Paleozoic
    C) Mesozoic
    D) Cenozoic

Science Online bookg.msscience.com/vocabulary_puzzlemaker

Thinking Critically

16. **Infer** why plants couldn't move onto land until an ozone layer formed.

17. **Discuss** why trilobites are classified as index fossils.

18. **Compare and contrast** the most significant difference between Precambrian life-forms and Paleozoic life-forms.

19. **Describe** how natural selection is related to organic evolution.

20. **Explain** In the early 1800s, a naturalist proposed that the giraffe species has a long neck as a result of years of stretching their necks to reach leaves in tall trees. Why isn't this true?

21. **Infer** Use the outlines of the present-day continents to make a sketch of Pangaea.

22. **Form Hypotheses** Suggest some reasons why trilobites might have become extinct at the end of the Paleozoic Era.

23. **Interpret Data** A student found what she thought was a piece of dinosaur bone in Pleistocene sediment. How likely is it that she is right? Explain.

24. **Infer** why mammals didn't become dominant until after the dinosaurs disappeared.

Performance Activities

25. **Make a Model** In the Section 2 Lab, you learned how a particular characteristic might evolve within a species. Modify the experimental model by using color instead of height as a characteristic. Design your activity with the understanding that varimals live in a dark-colored forest environment.

26. **Make a Display** Certain groups of animals have dominated the land throughout geologic time. Use your textbook and other references to discover some of the dominant species of each era. Make a display that illustrates some animals from each era. Be sure to include appropriate habitats.

Applying Math

Use the graph below to answer questions 27 and 28.

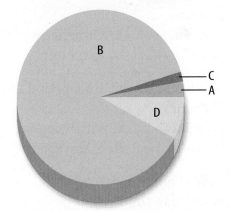

27. **Modeling Geologic Time** The circle graph above represents geologic time. Determine which interval of geologic time is represented by each portion of the graph. Which interval was longest? Which do we know the least about? Which of these intervals is getting larger?

28. **Interpret Data** The Cenozoic Era has lasted 65 million years. What percentage of Earth's 4.5-billion-year history is that?

Part 1 | Multiple Choice

Record your answers on the answer sheet provided by your teacher or on a sheet of paper.

Examine the diagram below. Then answer questions 1–3.

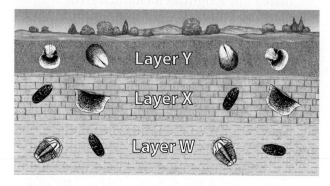

1. During which geologic time period did layer W form?
  A. Cambrian   **C.** Devonian
  B. Ordovician   **D.** Silurian

2. During which geologic time period did layer X form?
  A. Devonian   **C.** Ordovician
  B. Silurian   **D.** Cambrian

3. During which geologic time period did layer Y form?
  A. Cambrian   **C.** Mississippian
  B. Silurian   **D.** Ordovician

4. When did dinosaurs roam Earth?
  A. Precambrian time
  B. Paleozoic Era
  C. Mesozoic Era
  D. Cenozoic Era

5. What is the name of the supercontinent that formed at the end of the Paleozoic Era?
  A. Gondwanaland
  B. Eurasia
  C. Laurasia
  D. Pangaea

6. During which geologic period did modern humans evolve?
  A. Quaternary
  B. Triassic
  C. Ordovician
  D. Tertiary

7. How many body lobes did trilobites have?
  A. one   **C.** three
  B. two   **D.** four

8. Which mountain range formed because India collided with Asia?
  A. Alps   **C.** Ural
  B. Andes   **D.** Himalaya

Use the diagram below to answer questions 9–11.

	Quaternary Period	Holocene Epoch
		Pleistocene Epoch
Cenozoic Era	Tertiary Period	Pliocene Epoch
		Miocene Epoch
		Oligocene Epoch
		Eocene Epoch
		Paleocene Epoch

9. What is the oldest epoch in the Cenozoic Era?
  A. Pleistocene   **C.** Miocene
  B. Paleocene   **D.** Holocene

10. What is the youngest epoch in the Cenozoic Era?
  A. Miocene   **C.** Paleocene
  B. Holocene   **D.** Eocene

11. Which epoch is part of the Quaternary Period?
  A. Oligocene   **C.** Pleistocene
  B. Eocene   **D.** Pliocene

Part 2 | Short Response/Grid In

Record your answers on the answer sheet provided by your teacher or on a sheet of paper.

12. Who was Charles Darwin? How did he contribute to science?

13. Explain one hypothesis about why dinosaurs might have become extinct.

14. Describe *Archaeopteryx*. Why is this an important fossil?

15. Why do many scientists think that dinosaurs were warm-blooded?

16. What are stromatolites? How do they form?

17. Define the term *species*.

Select one of the equations below to help you answer questions 18–20.

$$\text{time} = \text{distance} \div \text{speed}$$

or

$$\text{speed} = \text{distance} \div \text{time}$$

18. It recently was estimated that *T. rex* could run no faster than about 11 m/s. At this speed, how long would it take *T. rex* to run 200 m?

19. A typical ornithopod (plant-eating dinosaur that walked on two legs) probably moved at a speed of about 2 m/s. How long would it take this dinosaur to run 200 m?

20. In 1996, Michael Johnson ran 200 m in 19.32 s. What was his average speed? How does this compare with *T. rex*?

Test-Taking Tip

Show Your Work For constructed response questions, show all of your work and any calculations on your answer sheet.

Part 3 | Open Ended

Record your answers on a sheet of paper.

Use the diagram below to answer questions 21 and 22. It shows the time ranges of various types of organisms on Earth. When a bar is wider, there were more species of that type of organism (higher diversity).

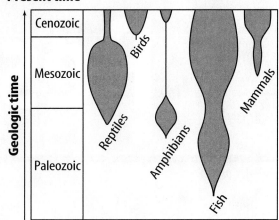

21. Describe how the diversity of reptiles changed through time.

22. How has the diversity of mammals changed through time? Do you see any relationship with how reptile diversity has changed?

23. Why might mammals in Australia be so much different than mammals on other continents?

24. Describe how natural selection might cause a species to change through time.

25. How did early photosynthetic organisms change the conditions on Earth to allow more advanced organisms to flourish?

26. What are mass extinctions? How have they affected life on Earth?

27. Write a description of what Earth was like during Precambrian time. Summarize how Earth was different than it is now.

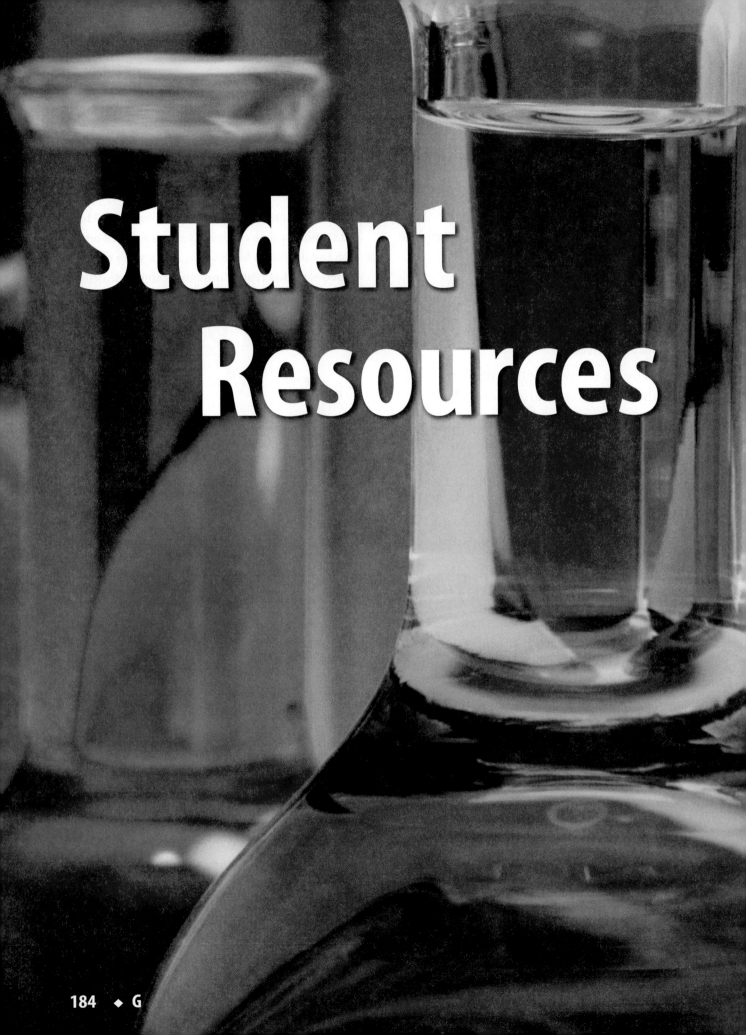

Student Resources

CONTENTS

Scientific Methods

Scientists use an orderly approach called the scientific method to solve problems. This includes organizing and recording data so others can understand them. Scientists use many variations in this method when they solve problems.

Identify a Question

The first step in a scientific investigation or experiment is to identify a question to be answered or a problem to be solved. For example, you might ask which gasoline is the most efficient.

Gather and Organize Information

After you have identified your question, begin gathering and organizing information. There are many ways to gather information, such as researching in a library, interviewing those knowledgeable about the subject, testing and working in the laboratory and field. Fieldwork is investigations and observations done outside of a laboratory.

Researching Information Before moving in a new direction, it is important to gather the information that already is known about the subject. Start by asking yourself questions to determine exactly what you need to know. Then you will look for the information in various reference sources, like the student is doing in **Figure 1.** Some sources may include textbooks, encyclopedias, government documents, professional journals, science magazines, and the Internet. Always list the sources of your information.

Figure 1 The Internet can be a valuable research tool.

Evaluate Sources of Information Not all sources of information are reliable. You should evaluate all of your sources of information, and use only those you know to be dependable. For example, if you are researching ways to make homes more energy efficient, a site written by the U.S. Department of Energy would be more reliable than a site written by a company that is trying to sell a new type of weatherproofing material. Also, remember that research always is changing. Consult the most current resources available to you. For example, a 1985 resource about saving energy would not reflect the most recent findings.

Sometimes scientists use data that they did not collect themselves, or conclusions drawn by other researchers. This data must be evaluated carefully. Ask questions about how the data were obtained, if the investigation was carried out properly, and if it has been duplicated exactly with the same results. Would you reach the same conclusion from the data? Only when you have confidence in the data can you believe it is true and feel comfortable using it.

Interpret Scientific Illustrations As you research a topic in science, you will see drawings, diagrams, and photographs to help you understand what you read. Some illustrations are included to help you understand an idea that you can't see easily by yourself, like the tiny particles in an atom in **Figure 2.** A drawing helps many people to remember details more easily and provides examples that clarify difficult concepts or give additional information about the topic you are studying. Most illustrations have labels or a caption to identify or to provide more information.

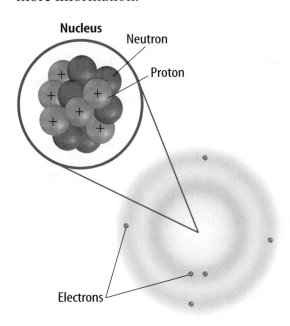

Figure 2 This drawing shows an atom of carbon with its six protons, six neutrons, and six electrons.

Concept Maps One way to organize data is to draw a diagram that shows relationships among ideas (or concepts). A concept map can help make the meanings of ideas and terms more clear, and help you understand and remember what you are studying. Concept maps are useful for breaking large concepts down into smaller parts, making learning easier.

Network Tree A type of concept map that not only shows a relationship, but how the concepts are related is a network tree, shown in **Figure 3.** In a network tree, the words are written in the ovals, while the description of the type of relationship is written across the connecting lines.

When constructing a network tree, write down the topic and all major topics on separate pieces of paper or notecards. Then arrange them in order from general to specific. Branch the related concepts from the major concept and describe the relationship on the connecting line. Continue to more specific concepts until finished.

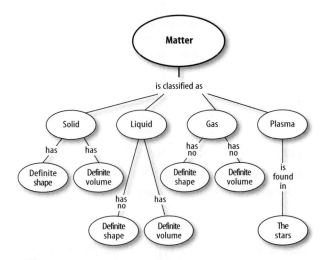

Figure 3 A network tree shows how concepts or objects are related.

Events Chain Another type of concept map is an events chain. Sometimes called a flow chart, it models the order or sequence of items. An events chain can be used to describe a sequence of events, the steps in a procedure, or the stages of a process.

When making an events chain, first find the one event that starts the chain. This event is called the initiating event. Then, find the next event and continue until the outcome is reached, as shown in **Figure 4.**

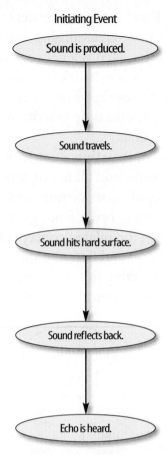

Figure 4 Events-chain concept maps show the order of steps in a process or event. This concept map shows how a sound makes an echo.

Cycle Map A specific type of events chain is a cycle map. It is used when the series of events do not produce a final outcome, but instead relate back to the beginning event, such as in **Figure 5.** Therefore, the cycle repeats itself.

To make a cycle map, first decide what event is the beginning event. This is also called the initiating event. Then list the next events in the order that they occur, with the last event relating back to the initiating event. Words can be written between the events that describe what happens from one event to the next. The number of events in a cycle map can vary, but usually contain three or more events.

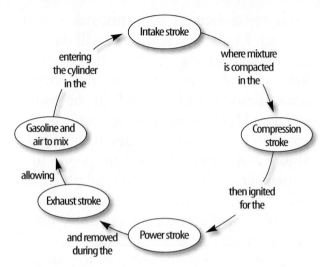

Figure 5 A cycle map shows events that occur in a cycle.

Spider Map A type of concept map that you can use for brainstorming is the spider map. When you have a central idea, you might find that you have a jumble of ideas that relate to it but are not necessarily clearly related to each other. The spider map on sound in **Figure 6** shows that if you write these ideas outside the main concept, then you can begin to separate and group unrelated terms so they become more useful.

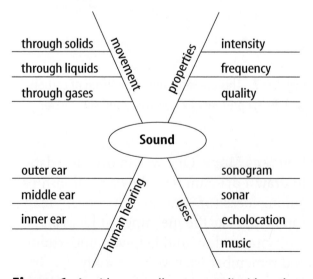

Figure 6 A spider map allows you to list ideas that relate to a central topic but not necessarily to one another.

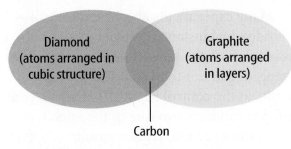

Figure 7 This Venn diagram compares and contrasts two substances made from carbon.

Venn Diagram To illustrate how two subjects compare and contrast you can use a Venn diagram. You can see the characteristics that the subjects have in common and those that they do not, shown in **Figure 7.**

To create a Venn diagram, draw two overlapping ovals that that are big enough to write in. List the characteristics unique to one subject in one oval, and the characteristics of the other subject in the other oval. The characteristics in common are listed in the overlapping section.

Make and Use Tables One way to organize information so it is easier to understand is to use a table. Tables can contain numbers, words, or both.

To make a table, list the items to be compared in the first column and the characteristics to be compared in the first row. The title should clearly indicate the content of the table, and the column or row heads should be clear. Notice that in **Table 1** the units are included.

Table 1 Recyclables Collected During Week			
Day of Week	Paper (kg)	Aluminum (kg)	Glass (kg)
Monday	5.0	4.0	12.0
Wednesday	4.0	1.0	10.0
Friday	2.5	2.0	10.0

Make a Model One way to help you better understand the parts of a structure, the way a process works, or to show things too large or small for viewing is to make a model. For example, an atomic model made of a plastic-ball nucleus and pipe-cleaner electron shells can help you visualize how the parts of an atom relate to each other. Other types of models can by devised on a computer or represented by equations.

Form a Hypothesis

A possible explanation based on previous knowledge and observations is called a hypothesis. After researching gasoline types and recalling previous experiences in your family's car you form a hypothesis—our car runs more efficiently because we use premium gasoline. To be valid, a hypothesis has to be something you can test by using an investigation.

Predict When you apply a hypothesis to a specific situation, you predict something about that situation. A prediction makes a statement in advance, based on prior observation, experience, or scientific reasoning. People use predictions to make everyday decisions. Scientists test predictions by performing investigations. Based on previous observations and experiences, you might form a prediction that cars are more efficient with premium gasoline. The prediction can be tested in an investigation.

Design an Experiment A scientist needs to make many decisions before beginning an investigation. Some of these include: how to carry out the investigation, what steps to follow, how to record the data, and how the investigation will answer the question. It also is important to address any safety concerns.

Test the Hypothesis

Now that you have formed your hypothesis, you need to test it. Using an investigation, you will make observations and collect data, or information. This data might either support or not support your hypothesis. Scientists collect and organize data as numbers and descriptions.

Follow a Procedure In order to know what materials to use, as well as how and in what order to use them, you must follow a procedure. **Figure 8** shows a procedure you might follow to test your hypothesis.

Procedure

1. Use regular gasoline for two weeks.
2. Record the number of kilometers between fill-ups and the amount of gasoline used.
3. Switch to premium gasoline for two weeks.
4. Record the number of kilometers between fill-ups and the amount of gasoline used.

Figure 8 A procedure tells you what to do step by step.

Identify and Manipulate Variables and Controls In any experiment, it is important to keep everything the same except for the item you are testing. The one factor you change is called the independent variable. The change that results is the dependent variable. Make sure you have only one independent variable, to assure yourself of the cause of the changes you observe in the dependent variable. For example, in your gasoline experiment the type of fuel is the independent variable. The dependent variable is the efficiency.

Many experiments also have a control—an individual instance or experimental subject for which the independent variable is not changed. You can then compare the test results to the control results. To design a control you can have two cars of the same type. The control car uses regular gasoline for four weeks. After you are done with the test, you can compare the experimental results to the control results.

Collect Data

Whether you are carrying out an investigation or a short observational experiment, you will collect data, as shown in **Figure 9.** Scientists collect data as numbers and descriptions and organize it in specific ways.

Observe Scientists observe items and events, then record what they see. When they use only words to describe an observation, it is called qualitative data. Scientists' observations also can describe how much there is of something. These observations use numbers, as well as words, in the description and are called quantitative data. For example, if a sample of the element gold is described as being "shiny and very dense" the data are qualitative. Quantitative data on this sample of gold might include "a mass of 30 g and a density of 19.3 g/cm^3."

Figure 9 Collecting data is one way to gather information directly.

Figure 10 Record data neatly and clearly so it is easy to understand.

When you make observations you should examine the entire object or situation first, and then look carefully for details. It is important to record observations accurately and completely. Always record your notes immediately as you make them, so you do not miss details or make a mistake when recording results from memory. Never put unidentified observations on scraps of paper. Instead they should be recorded in a notebook, like the one in **Figure 10.** Write your data neatly so you can easily read it later. At each point in the experiment, record your observations and label them. That way, you will not have to determine what the figures mean when you look at your notes later. Set up any tables that you will need to use ahead of time, so you can record any observations right away. Remember to avoid bias when collecting data by not including personal thoughts when you record observations. Record only what you observe.

Estimate Scientific work also involves estimating. To estimate is to make a judgment about the size or the number of something without measuring or counting. This is important when the number or size of an object or population is too large or too difficult to accurately count or measure.

Sample Scientists may use a sample or a portion of the total number as a type of estimation. To sample is to take a small, representative portion of the objects or organisms of a population for research. By making careful observations or manipulating variables within that portion of the group, information is discovered and conclusions are drawn that might apply to the whole population. A poorly chosen sample can be unrepresentative of the whole. If you were trying to determine the rainfall in an area, it would not be best to take a rainfall sample from under a tree.

Measure You use measurements everyday. Scientists also take measurements when collecting data. When taking measurements, it is important to know how to use measuring tools properly. Accuracy also is important.

Length To measure length, the distance between two points, scientists use meters. Smaller measurements might be measured in centimeters or millimeters.

Length is measured using a metric ruler or meter stick. When using a metric ruler, line up the 0-cm mark with the end of the object being measured and read the number of the unit where the object ends. Look at the metric ruler shown in **Figure 11.** The centimeter lines are the long, numbered lines, and the shorter lines are millimeter lines. In this instance, the length would be 4.50 cm.

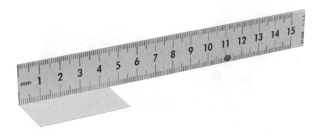

Figure 11 This metric ruler has centimeter and millimeter divisions.

Science Skill Handbook (sidebar)

Mass The SI unit for mass is the kilogram (kg). Scientists can measure mass using units formed by adding metric prefixes to the unit gram (g), such as milligram (mg). To measure mass, you might use a triple-beam balance similar to the one shown in **Figure 12.** The balance has a pan on one side and a set of beams on the other side. Each beam has a rider that slides on the beam.

When using a triple-beam balance, place an object on the pan. Slide the largest rider along its beam until the pointer drops below zero. Then move it back one notch. Repeat the process for each rider proceeding from the larger to smaller until the pointer swings an equal distance above and below the zero point. Sum the masses on each beam to find the mass of the object. Move all riders back to zero when finished.

Instead of putting materials directly on the balance, scientists often take a tare of a container. A tare is the mass of a container into which objects or substances are placed for measuring their masses. To mass objects or substances, find the mass of a clean container. Remove the container from the pan, and place the object or substances in the container. Find the mass of the container with the materials in it. Subtract the mass of the empty container from the mass of the filled container to find the mass of the materials you are using.

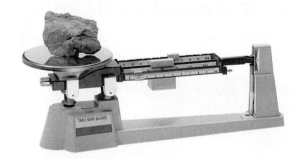

Figure 12 A triple-beam balance is used to determine the mass of an object.

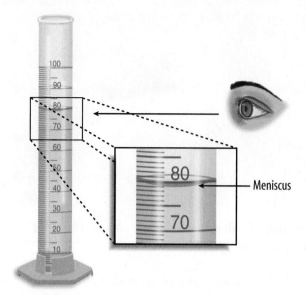

Meniscus

Figure 13 Graduated cylinders measure liquid volume.

Liquid Volume To measure liquids, the unit used is the liter. When a smaller unit is needed, scientists might use a milliliter. Because a milliliter takes up the volume of a cube measuring 1 cm on each side it also can be called a cubic centimeter ($cm^3 = cm \times cm \times cm$).

You can use beakers and graduated cylinders to measure liquid volume. A graduated cylinder, shown in **Figure 13,** is marked from bottom to top in milliliters. In lab, you might use a 10-mL graduated cylinder or a 100-mL graduated cylinder. When measuring liquids, notice that the liquid has a curved surface. Look at the surface at eye level, and measure the bottom of the curve. This is called the meniscus. The graduated cylinder in **Figure 13** contains 79.0 mL, or 79.0 cm^3, of a liquid.

Temperature Scientists often measure temperature using the Celsius scale. Pure water has a freezing point of 0°C and boiling point of 100°C. The unit of measurement is degrees Celsius. Two other scales often used are the Fahrenheit and Kelvin scales.

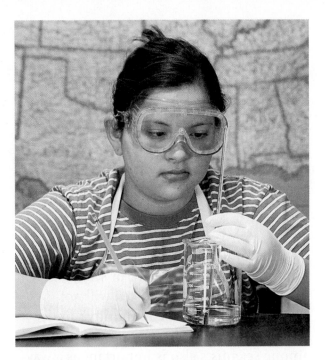

Figure 14 A thermometer measures the temperature of an object.

Scientists use a thermometer to measure temperature. Most thermometers in a laboratory are glass tubes with a bulb at the bottom end containing a liquid such as colored alcohol. The liquid rises or falls with a change in temperature. To read a glass thermometer like the thermometer in **Figure 14,** rotate it slowly until a red line appears. Read the temperature where the red line ends.

Form Operational Definitions An operational definition defines an object by how it functions, works, or behaves. For example, when you are playing hide and seek and a tree is home base, you have created an operational definition for a tree.

Objects can have more than one operational definition. For example, a ruler can be defined as a tool that measures the length of an object (how it is used). It can also be a tool with a series of marks used as a standard when measuring (how it works).

Analyze the Data

To determine the meaning of your observations and investigation results, you will need to look for patterns in the data. Then you must think critically to determine what the data mean. Scientists use several approaches when they analyze the data they have collected and recorded. Each approach is useful for identifying specific patterns.

Interpret Data The word *interpret* means "to explain the meaning of something." When analyzing data from an experiment, try to find out what the data show. Identify the control group and the test group to see whether or not changes in the independent variable have had an effect. Look for differences in the dependent variable between the control and test groups.

Classify Sorting objects or events into groups based on common features is called classifying. When classifying, first observe the objects or events to be classified. Then select one feature that is shared by some members in the group, but not by all. Place those members that share that feature in a subgroup. You can classify members into smaller and smaller subgroups based on characteristics. Remember that when you classify, you are grouping objects or events for a purpose. Keep your purpose in mind as you select the features to form groups and subgroups.

Compare and Contrast Observations can be analyzed by noting the similarities and differences between two more objects or events that you observe. When you look at objects or events to see how they are similar, you are comparing them. Contrasting is looking for differences in objects or events.

Recognize Cause and Effect A cause is a reason for an action or condition. The effect is that action or condition. When two events happen together, it is not necessarily true that one event caused the other. Scientists must design a controlled investigation to recognize the exact cause and effect.

Draw Conclusions

When scientists have analyzed the data they collected, they proceed to draw conclusions about the data. These conclusions are sometimes stated in words similar to the hypothesis that you formed earlier. They may confirm a hypothesis, or lead you to a new hypothesis.

Infer Scientists often make inferences based on their observations. An inference is an attempt to explain observations or to indicate a cause. An inference is not a fact, but a logical conclusion that needs further investigation. For example, you may infer that a fire has caused smoke. Until you investigate, however, you do not know for sure.

Apply When you draw a conclusion, you must apply those conclusions to determine whether the data supports the hypothesis. If your data do not support your hypothesis, it does not mean that the hypothesis is wrong. It means only that the result of the investigation did not support the hypothesis. Maybe the experiment needs to be redesigned, or some of the initial observations on which the hypothesis was based were incomplete or biased. Perhaps more observation or research is needed to refine your hypothesis. A successful investigation does not always come out the way you originally predicted.

Avoid Bias Sometimes a scientific investigation involves making judgments. When you make a judgment, you form an opinion. It is important to be honest and not to allow any expectations of results to bias your judgments. This is important throughout the entire investigation, from researching to collecting data to drawing conclusions.

Communicate

The communication of ideas is an important part of the work of scientists. A discovery that is not reported will not advance the scientific community's understanding or knowledge. Communication among scientists also is important as a way of improving their investigations.

Scientists communicate in many ways, from writing articles in journals and magazines that explain their investigations and experiments, to announcing important discoveries on television and radio. Scientists also share ideas with colleagues on the Internet or present them as lectures, like the student is doing in **Figure 15.**

Figure 15 A student communicates to his peers about his investigation.

SAFETY SYMBOLS

SAFETY SYMBOLS	HAZARD	EXAMPLES	PRECAUTION	REMEDY
DISPOSAL	Special disposal procedures need to be followed.	certain chemicals, living organisms	Do not dispose of these materials in the sink or trash can.	Dispose of wastes as directed by your teacher.
BIOLOGICAL	Organisms or other biological materials that might be harmful to humans	bacteria, fungi, blood, unpreserved tissues, plant materials	Avoid skin contact with these materials. Wear mask or gloves.	Notify your teacher if you suspect contact with material. Wash hands thoroughly.
EXTREME TEMPERATURE	Objects that can burn skin by being too cold or too hot	boiling liquids, hot plates, dry ice, liquid nitrogen	Use proper protection when handling.	Go to your teacher for first aid.
SHARP OBJECT	Use of tools or glassware that can easily puncture or slice skin	razor blades, pins, scalpels, pointed tools, dissecting probes, broken glass	Practice common-sense behavior and follow guidelines for use of the tool.	Go to your teacher for first aid.
FUME	Possible danger to respiratory tract from fumes	ammonia, acetone, nail polish remover, heated sulfur, moth balls	Make sure there is good ventilation. Never smell fumes directly. Wear a mask.	Leave foul area and notify your teacher immediately.
ELECTRICAL	Possible danger from electrical shock or burn	improper grounding, liquid spills, short circuits, exposed wires	Double-check setup with teacher. Check condition of wires and apparatus.	Do not attempt to fix electrical problems. Notify your teacher immediately.
IRRITANT	Substances that can irritate the skin or mucous membranes of the respiratory tract	pollen, moth balls, steel wool, fiberglass, potassium permanganate	Wear dust mask and gloves. Practice extra care when handling these materials.	Go to your teacher for first aid.
CHEMICAL	Chemicals can react with and destroy tissue and other materials	bleaches such as hydrogen peroxide; acids such as sulfuric acid, hydrochloric acid; bases such as ammonia, sodium hydroxide	Wear goggles, gloves, and an apron.	Immediately flush the affected area with water and notify your teacher.
TOXIC	Substance may be poisonous if touched, inhaled, or swallowed.	mercury, many metal compounds, iodine, poinsettia plant parts	Follow your teacher's instructions.	Always wash hands thoroughly after use. Go to your teacher for first aid.
FLAMMABLE	Flammable chemicals may be ignited by open flame, spark, or exposed heat.	alcohol, kerosene, potassium permanganate	Avoid open flames and heat when using flammable chemicals.	Notify your teacher immediately. Use fire safety equipment if applicable.
OPEN FLAME	Open flame in use, may cause fire.	hair, clothing, paper, synthetic materials	Tie back hair and loose clothing. Follow teacher's instruction on lighting and extinguishing flames.	Notify your teacher immediately. Use fire safety equipment if applicable.

 Eye Safety
Proper eye protection should be worn at all times by anyone performing or observing science activities.

 Clothing Protection
This symbol appears when substances could stain or burn clothing.

 Animal Safety
This symbol appears when safety of animals and students must be ensured.

 Handwashing
After the lab, wash hands with soap and water before removing goggles.

Safety in the Science Laboratory

The science laboratory is a safe place to work if you follow standard safety procedures. Being responsible for your own safety helps to make the entire laboratory a safer place for everyone. When performing any lab, read and apply the caution statements and safety symbol listed at the beginning of the lab.

General Safety Rules

1. Obtain your teacher's permission to begin all investigations and use laboratory equipment.

2. Study the procedure. Ask your teacher any questions. Be sure you understand safety symbols shown on the page.

3. Notify your teacher about allergies or other health conditions which can affect your participation in a lab.

4. Learn and follow use and safety procedures for your equipment. If unsure, ask your teacher.

5. Never eat, drink, chew gum, apply cosmetics, or do any personal grooming in the lab. Never use lab glassware as food or drink containers. Keep your hands away from your face and mouth.

6. Know the location and proper use of the safety shower, eye wash, fire blanket, and fire alarm.

Prevent Accidents

1. Use the safety equipment provided to you. Goggles and a safety apron should be worn during investigations.

2. Do NOT use hair spray, mousse, or other flammable hair products. Tie back long hair and tie down loose clothing.

3. Do NOT wear sandals or other open-toed shoes in the lab.

4. Remove jewelry on hands and wrists. Loose jewelry, such as chains and long necklaces, should be removed to prevent them from getting caught in equipment.

5. Do not taste any substances or draw any material into a tube with your mouth.

6. Proper behavior is expected in the lab. Practical jokes and fooling around can lead to accidents and injury.

7. Keep your work area uncluttered.

Laboratory Work

1. Collect and carry all equipment and materials to your work area before beginning a lab.

2. Remain in your own work area unless given permission by your teacher to leave it.

3. Always slant test tubes away from yourself and others when heating them, adding substances to them, or rinsing them.

4. If instructed to smell a substance in a container, hold the container a short distance away and fan vapors towards your nose.

5. Do NOT substitute other chemicals/substances for those in the materials list unless instructed to do so by your teacher.

6. Do NOT take any materials or chemicals outside of the laboratory.

7. Stay out of storage areas unless instructed to be there and supervised by your teacher.

Laboratory Cleanup

1. Turn off all burners, water, and gas, and disconnect all electrical devices.

2. Clean all pieces of equipment and return all materials to their proper places.

3. Dispose of chemicals and other materials as directed by your teacher. Place broken glass and solid substances in the proper containers. Never discard materials in the sink.

4. Clean your work area.

5. Wash your hands with soap and water thoroughly BEFORE removing your goggles.

Emergencies

1. Report any fire, electrical shock, glassware breakage, spill, or injury, no matter how small, to your teacher immediately. Follow his or her instructions.

2. If your clothing should catch fire, STOP, DROP, and ROLL. If possible, smother it with the fire blanket or get under a safety shower. NEVER RUN.

3. If a fire should occur, turn off all gas and leave the room according to established procedures.

4. In most instances, your teacher will clean up spills. Do NOT attempt to clean up spills unless you are given permission and instructions to do so.

5. If chemicals come into contact with your eyes or skin, notify your teacher immediately. Use the eyewash or flush your skin or eyes with large quantities of water.

6. The fire extinguisher and first-aid kit should only be used by your teacher unless it is an extreme emergency and you have been given permission.

7. If someone is injured or becomes ill, only a professional medical provider or someone certified in first aid should perform first-aid procedures.

EXTRA Try at Home Labs

From Your Kitchen, Junk Drawer, or Yard

1 3-D Maps

▶ **Real-World Question**

How can you make a topographic map of a room in your house?

Possible Materials
- meterstick
- metric ruler
- metric tape measure
- poster board
- black marker
- construction paper
- transparent tape

▶ **Procedure**

1. Measure the length and width of your room in meters. Include the measurements of any odd shapes or angles in the room, as well as elevations.
2. Decide upon a scale for your map.
3. Using your scale, draw the outline of your room on the poster board.
4. Measure the length, width, and height of a piece of furniture.
5. Using your scale, measure and cut out the sides for a model of the furniture piece from construction paper. Tape the pieces of the model together.
6. Place your furniture model on your map to match the actual piece's location in your room.
7. Construct two or three other models of furniture for your map.

▶ **Conclude and Apply**

1. What scale did you use for your map?
2. Infer how a biologist might use a topographic map.

2 Rock and Roll

▶ **Real-World Question**

How can we model the weathering of rock?

Possible Materials
- white glue
- sand
- plastic bowl
- plastic spoon
- cookie tray
- barbecue brush
- aluminum foil
- cooking oil
- empty coffee can with lid
- water
- measuring cup
- transparent packing tape

▶ **Procedure**

1. Make your own sedimentary rocks by adding equal amounts of white glue and sand to a bowl. Stir the sand and glue together until you make several small lumps.
2. Lay aluminum foil on the bottom of a cookie tray and coat the foil with cooking oil.
3. Lay your rocks on the tray in direct sunlight for three days until they dry.
4. Place your rocks in a coffee can and pour 50 mL of water into the can.
5. Place the lid on the can and secure the lid with thick transparent tape.
6. Shake the contents of the can for 4 min, open the lid, and observe your rocks.

▶ **Conclude and Apply**

1. Describe what happened to your sedimentary rocks.
2. Infer how this activity modeled the weathering of rocks.

Adult supervision required for all labs.

3 Modeling Mudslides

Real-World Question
How can mudslides be prevented?

Possible Materials
- deep, rectangular basin
- bowl
- measuring cup
- hose
- water
- soil (potting or garden)
- blocks of wood or bricks
- protractor
- sod
- several houseplants

Procedure
1. Pour soil into one-half of a deep, rectangular basin. Lightly pack the soil down.
2. Prop the container up on wooden blocks until the end with the soil in it is raised to a 60° angle.
3. Fill a bowl with water and slowly pour the water over the soil near the edge of the basin.
4. Continue pouring water onto the soil until a mudslide is created.
5. Clean the basin and repeat the lab, but plant sod or houseplants in the soil before watering it.

Conclude and Apply
1. Describe what happened to the soil without plants and with plants.
2. Explain how this lab modeled a mudslide.
3. Describe the relationship between vegetation and mudslides.

4 In Deep Water

Real-World Question
How can the water table and a well be modeled?

Possible Materials
- clear-plastic drink bottle (500-mL)
- aquarium gravel
- water
- blue food dye
- measuring cup
- long dropper

Procedure
1. Fill a clear-plastic bottle with aquarium gravel.
2. Pour 450 mL of water into the the measuring cup and add several drops of blue food dye.
3. Pour 300 mL of the blue water into the bottle with the gravel. Insert the dropper down into the gravel and try to suck out some of the water.
4. Pour another 150 mL of water into the bottle and try to suck out some of the water with the dropper.

Conclude and Apply
1. Describe how this lab models the water table and a well.
2. Infer how deep a well must be dug for it to yield water.
3. Infer why some wells only yield water at certain times of the year.

Extra Try at Home Labs

5 Making Burrows

▶ *Real-World Question*

How does burrowing affect sediment layers?

Possible Materials 🖼️
- clear-glass bowl
- white flour
- colored gelatin powder (3 packages)
- paintbrush
- pencil

▶ *Procedure*

1. Add 3 cm of white flour to the bowl. Flatten the top of the flour layer.
2. Carefully sprinkle gelatin powder over the flour to form a colored layer about 0.25 cm thick.
3. The two layers represent two different layers of sediment.
4. Use a paintbrush or pencil to make "burrows" in the "sediment."
5. Make sure to make some of the burrows at the edge of the bowl so that you can see how it affects the sediment.
6. Continue to make more burrows and observe the effect on the two layers.

▶ *Conclude and Apply*

1. How did the two layers of powder change as you continued to make burrows?
2. Were the "trace fossils" easy to recognize at first? How about after a lot of burrowing?
3. How do you think burrowing animals affect layers of sediment on the ocean floor? How could this burrowing be recognized in rock?

6 History in a Bottle

▶ *Real-World Question*

What does the geologic column look like?

Possible Materials 🖼️ 🖼️
- clear-plastic 2-liter bottle
- scissors
- 3-in × 5-in index cards
- colored markers
- permanent marker
- transparent tape
- metric ruler
- sand (3 different colors)
- aquarium gravel (3 different colors)

▶ *Procedure*

1. Cut the top 5 cm off a clear-plastic 2-L soda bottle. Remove the label.
2. Cut 12 square cards measuring 2 cm × 2 cm.
3. Draw a picture of a trilobite, coral, fish, amphibian, insect, reptile, mouse, coniferous tree, dinosaur, bird, flower, large mammal, and human on the 12 cards.
4. Starting at the bottom, inside of the bottle and working up, tape the trilobite, coral, fish, amphibian, insect, and reptile pictures face out in that order. The reptile picture should be about half way up the bottle.
5. Pour red sand into the bottle until it covers your reptile picture.
6. Tape the mouse, conifer tree, dinosaur, bird, and flower pictures above the red sand in that order. Pour in blue sand until it covers the flower picture.
7. Tape the large mammal and human pictures above the blue sand. Pour green sand into the bottle until it covers the person.

▶ *Conclude and Apply*

1. Research what era each color of sand represents.
2. Infer why few fossils of organisms living before the Paleozoic Era are found.

Adult supervision required for all labs.

Computer Skills

People who study science rely on computers, like the one in **Figure 16,** to record and store data and to analyze results from investigations. Whether you work in a laboratory or just need to write a lab report with tables, good computer skills are a necessity.

Using the computer comes with responsibility. Issues of ownership, security, and privacy can arise. Remember, if you did not author the information you are using, you must provide a source for your information. Also, anything on a computer can be accessed by others. Do not put anything on the computer that you would not want everyone to know. To add more security to your work, use a password.

Use a Word Processing Program

A computer program that allows you to type your information, change it as many times as you need to, and then print it out is called a word processing program. Word processing programs also can be used to make tables.

Figure 16 A computer will make reports neater and more professional looking.

Learn the Skill To start your word processing program, a blank document, sometimes called "Document 1," appears on the screen. To begin, start typing. To create a new document, click the *New* button on the standard tool bar. These tips will help you format the document.

- The program will automatically move to the next line; press *Enter* if you wish to start a new paragraph.
- Symbols, called non-printing characters, can be hidden by clicking the *Show/Hide* button on your toolbar.
- To insert text, move the cursor to the point where you want the insertion to go, click on the mouse once, and type the text.
- To move several lines of text, select the text and click the *Cut* button on your toolbar. Then position your cursor in the location that you want to move the cut text and click *Paste.* If you move to the wrong place, click *Undo.*
- The spell check feature does not catch words that are misspelled to look like other words, like "cold" instead of "gold." Always reread your document to catch all spelling mistakes.
- To learn about other word processing methods, read the user's manual or click on the *Help* button.
- You can integrate databases, graphics, and spreadsheets into documents by copying from another program and pasting it into your document, or by using desktop publishing (DTP). DTP software allows you to put text and graphics together to finish your document with a professional look. This software varies in how it is used and its capabilities.

Use a Database

A collection of facts stored in a computer and sorted into different fields is called a database. A database can be reorganized in any way that suits your needs.

Learn the Skill A computer program that allows you to create your own database is a database management system (DBMS). It allows you to add, delete, or change information. Take time to get to know the features of your database software.

- Determine what facts you would like to include and research to collect your information.
- Determine how you want to organize the information.
- Follow the instructions for your particular DBMS to set up fields. Then enter each item of data in the appropriate field.
- Follow the instructions to sort the information in order of importance.
- Evaluate the information in your database, and add, delete, or change as necessary.

Use the Internet

The Internet is a global network of computers where information is stored and shared. To use the Internet, like the students in **Figure 17,** you need a modem to connect your computer to a phone line and an Internet Service Provider account.

Learn the Skill To access internet sites and information, use a "Web browser," which lets you view and explore pages on the World Wide Web. Each page is its own site, and each site has its own address, called a URL. Once you have found a Web browser, follow these steps for a search (this also is how you search a database).

Figure 17 The Internet allows you to search a global network for a variety of information.

- Be as specific as possible. If you know you want to research "gold," don't type in "elements." Keep narrowing your search until you find what you want.
- Web sites that end in *.com* are commercial Web sites; *.org, .edu,* and *.gov* are non-profit, educational, or government Web sites.
- Electronic encyclopedias, almanacs, indexes, and catalogs will help locate and select relevant information.
- Develop a "home page" with relative ease. When developing a Web site, NEVER post pictures or disclose personal information such as location, names, or phone numbers. Your school or community usually can host your Web site. A basic understanding of HTML (hypertext mark-up language), the language of Web sites, is necessary. Software that creates HTML code is called authoring software, and can be downloaded free from many Web sites. This software allows text and pictures to be arranged as the software is writing the HTML code.

Use a Spreadsheet

A spreadsheet, shown in **Figure 18,** can perform mathematical functions with any data arranged in columns and rows. By entering a simple equation into a cell, the program can perform operations in specific cells, rows, or columns.

Learn the Skill Each column (vertical) is assigned a letter, and each row (horizontal) is assigned a number. Each point where a row and column intersect is called a cell, and is labeled according to where it is located—Column A, Row 1 (A1).

- Decide how to organize the data, and enter it in the correct row or column.
- Spreadsheets can use standard formulas or formulas can be customized to calculate cells.
- To make a change, click on a cell to make it activate, and enter the edited data or formula.
- Spreadsheets also can display your results in graphs. Choose the style of graph that best represents the data.

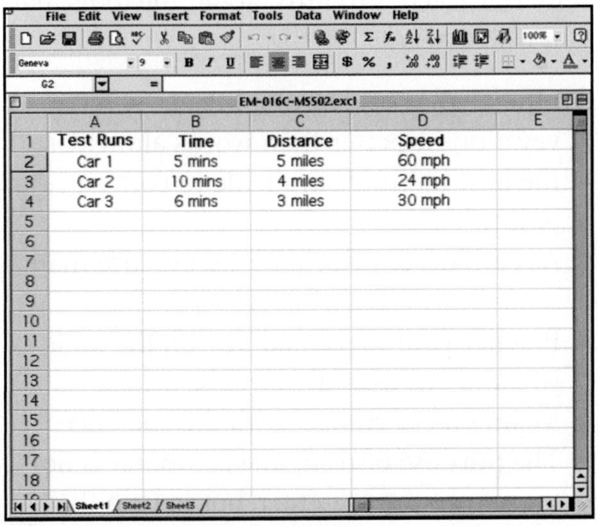

Figure 18 A spreadsheet allows you to perform mathematical operations on your data.

Use Graphics Software

Adding pictures, called graphics, to your documents is one way to make your documents more meaningful and exciting. This software adds, edits, and even constructs graphics. There is a variety of graphics software programs. The tools used for drawing can be a mouse, keyboard, or other specialized devices. Some graphics programs are simple. Others are complicated, called computer-aided design (CAD) software.

Learn the Skill It is important to have an understanding of the graphics software being used before starting. The better the software is understood, the better the results. The graphics can be placed in a word-processing document.

- Clip art can be found on a variety of internet sites, and on CDs. These images can be copied and pasted into your document.
- When beginning, try editing existing drawings, then work up to creating drawings.
- The images are made of tiny rectangles of color called pixels. Each pixel can be altered.
- Digital photography is another way to add images. The photographs in the memory of a digital camera can be downloaded into a computer, then edited and added to the document.
- Graphics software also can allow animation. The software allows drawings to have the appearance of movement by connecting basic drawings automatically. This is called in-betweening, or tweening.
- Remember to save often.

Presentation Skills

Develop Multimedia Presentations

Most presentations are more dynamic if they include diagrams, photographs, videos, or sound recordings, like the one shown in **Figure 19.** A multimedia presentation involves using stereos, overhead projectors, televisions, computers, and more.

Learn the Skill Decide the main points of your presentation, and what types of media would best illustrate those points.

- Make sure you know how to use the equipment you are working with.
- Practice the presentation using the equipment several times.
- Enlist the help of a classmate to push play or turn lights out for you. Be sure to practice your presentation with him or her.
- If possible, set up all of the equipment ahead of time, and make sure everything is working properly.

Figure 19 These students are engaging the audience using a variety of tools.

Computer Presentations

There are many different interactive computer programs that you can use to enhance your presentation. Most computers have a compact disc (CD) drive that can play both CDs and digital video discs (DVDs). Also, there is hardware to connect a regular CD, DVD, or VCR. These tools will enhance your presentation.

Another method of using the computer to aid in your presentation is to develop a slide show using a computer program. This can allow movement of visuals at the presenter's pace, and can allow for visuals to build on one another.

Learn the Skill In order to create multimedia presentations on a computer, you need to have certain tools. These may include traditional graphic tools and drawing programs, animation programs, and authoring systems that tie everything together. Your computer will tell you which tools it supports. The most important step is to learn about the tools that you will be using.

- Often, color and strong images will convey a point better than words alone. Use the best methods available to convey your point.
- As with other presentations, practice many times.
- Practice your presentation with the tools you and any assistants will be using.
- Maintain eye contact with the audience. The purpose of using the computer is not to prompt the presenter, but to help the audience understand the points of the presentation.

Math Review

Use Fractions

A fraction compares a part to a whole. In the fraction $\frac{2}{3}$, the 2 represents the part and is the numerator. The 3 represents the whole and is the denominator.

Reduce Fractions To reduce a fraction, you must find the largest factor that is common to both the numerator and the denominator, the greatest common factor (GCF). Divide both numbers by the GCF. The fraction has then been reduced, or it is in its simplest form.

Example Twelve of the 20 chemicals in the science lab are in powder form. What fraction of the chemicals used in the lab are in powder form?

Step 1 Write the fraction.

$$\frac{part}{whole} = \frac{12}{20}$$

Step 2 To find the GCF of the numerator and denominator, list all of the factors of each number.

Factors of 12: 1, 2, 3, 4, 6, 12 (the numbers that divide evenly into 12)

Factors of 20: 1, 2, 4, 5, 10, 20 (the numbers that divide evenly into 20)

Step 3 List the common factors.

1, 2, 4.

Step 4 Choose the greatest factor in the list.

The GCF of 12 and 20 is 4.

Step 5 Divide the numerator and denominator by the GCF.

$$\frac{12 \div 4}{20 \div 4} = \frac{3}{5}$$

In the lab, $\frac{3}{5}$ of the chemicals are in powder form.

Practice Problem At an amusement park, 66 of 90 rides have a height restriction. What fraction of the rides, in its simplest form, has a height restriction?

Add and Subtract Fractions To add or subtract fractions with the same denominator, add or subtract the numerators and write the sum or difference over the denominator. After finding the sum or difference, find the simplest form for your fraction.

Example 1 In the forest outside your house, $\frac{1}{8}$ of the animals are rabbits, $\frac{3}{8}$ are squirrels, and the remainder are birds and insects. How many are mammals?

Step 1 Add the numerators.

$$\frac{1}{8} + \frac{3}{8} = \frac{(1+3)}{8} = \frac{4}{8}$$

Step 2 Find the GCF.

$$\frac{4}{8} \ (GCF, 4)$$

Step 3 Divide the numerator and denominator by the GCF.

$$\frac{4}{4} = 1, \ \frac{8}{4} = 2$$

$\frac{1}{2}$ of the animals are mammals.

Example 2 If $\frac{7}{16}$ of the Earth is covered by freshwater, and $\frac{1}{16}$ of that is in glaciers, how much freshwater is not frozen?

Step 1 Subtract the numerators.

$$\frac{7}{16} - \frac{1}{16} = \frac{(7-1)}{16} = \frac{6}{16}$$

Step 2 Find the GCF.

$$\frac{6}{16} \ (GCF, 2)$$

Step 3 Divide the numerator and denominator by the GCF.

$$\frac{6}{2} = 3, \ \frac{16}{2} = 8$$

$\frac{3}{8}$ of the freshwater is not frozen.

Practice Problem A bicycle rider is going 15 km/h for $\frac{4}{9}$ of his ride, 10 km/h for $\frac{2}{9}$ of his ride, and 8 km/h for the remainder of the ride. How much of his ride is he going over 8 km/h?

Unlike Denominators To add or subtract fractions with unlike denominators, first find the least common denominator (LCD). This is the smallest number that is a common multiple of both denominators. Rename each fraction with the LCD, and then add or subtract. Find the simplest form if necessary.

Example 1 A chemist makes a paste that is $\frac{1}{2}$ table salt (NaCl), $\frac{1}{3}$ sugar ($C_6H_{12}O_6$), and the rest water (H_2O). How much of the paste is a solid?

Step 1 Find the LCD of the fractions.

$$\frac{1}{2} + \frac{1}{3} \quad \text{(LCD, 6)}$$

Step 2 Rename each numerator and each denominator with the LCD.

$$1 \times 3 = 3, \quad 2 \times 3 = 6$$
$$1 \times 2 = 2, \quad 3 \times 2 = 6$$

Step 3 Add the numerators.

$$\frac{3}{6} + \frac{2}{6} = \frac{(3 + 2)}{6} = \frac{5}{6}$$

$\frac{5}{6}$ of the paste is a solid.

Example 2 The average precipitation in Grand Junction, CO, is $\frac{7}{10}$ inch in November, and $\frac{3}{5}$ inch in December. What is the total average precipitation?

Step 1 Find the LCD of the fractions.

$$\frac{7}{10} + \frac{3}{5} \quad \text{(LCD, 10)}$$

Step 2 Rename each numerator and each denominator with the LCD.

$$7 \times 1 = 7, \quad 10 \times 1 = 10$$
$$3 \times 2 = 6, \quad 5 \times 2 = 10$$

Step 3 Add the numerators.

$$\frac{7}{10} + \frac{6}{10} = \frac{(7 + 6)}{10} = \frac{13}{10}$$

$\frac{13}{10}$ inches total precipitation, or $1\frac{3}{10}$ inches.

Practice Problem On an electric bill, about $\frac{1}{8}$ of the energy is from solar energy and about $\frac{1}{10}$ is from wind power. How much of the total bill is from solar energy and wind power combined?

Example 3 In your body, $\frac{7}{10}$ of your muscle contractions are involuntary (cardiac and smooth muscle tissue). Smooth muscle makes $\frac{3}{15}$ of your muscle contractions. How many of your muscle contractions are made by cardiac muscle?

Step 1 Find the LCD of the fractions.

$$\frac{7}{10} - \frac{3}{15} \quad \text{(LCD, 30)}$$

Step 2 Rename each numerator and each denominator with the LCD.

$$7 \times 3 = 21, \quad 10 \times 3 = 30$$
$$3 \times 2 = 6, \quad 15 \times 2 = 30$$

Step 3 Subtract the numerators.

$$\frac{21}{30} - \frac{6}{30} = \frac{(21 - 6)}{30} = \frac{15}{30}$$

Step 4 Find the GCF.

$$\frac{15}{30} \quad \text{(GCF, 15)}$$

$$\frac{1}{2}$$

$\frac{1}{2}$ of all muscle contractions are cardiac muscle.

Example 4 Tony wants to make cookies that call for $\frac{3}{4}$ of a cup of flour, but he only has $\frac{1}{3}$ of a cup. How much more flour does he need?

Step 1 Find the LCD of the fractions.

$$\frac{3}{4} - \frac{1}{3} \quad \text{(LCD, 12)}$$

Step 2 Rename each numerator and each denominator with the LCD.

$$3 \times 3 = 9, \quad 4 \times 3 = 12$$
$$1 \times 4 = 4, \quad 3 \times 4 = 12$$

Step 3 Subtract the numerators.

$$\frac{9}{12} - \frac{4}{12} = \frac{(9 - 4)}{12} = \frac{5}{12}$$

$\frac{5}{12}$ of a cup of flour.

Practice Problem Using the information provided to you in Example 3 above, determine how many muscle contractions are voluntary (skeletal muscle).

Multiply Fractions To multiply with fractions, multiply the numerators and multiply the denominators. Find the simplest form if necessary.

Example Multiply $\frac{3}{5}$ by $\frac{1}{3}$.

Step 1 Multiply the numerators and denominators.

$$\frac{3}{5} \times \frac{1}{3} = \frac{(3 \times 1)}{(5 \times 3)} = \frac{3}{15}$$

Step 2 Find the GCF.

$$\frac{3}{15} \quad (GCF, 3)$$

Step 3 Divide the numerator and denominator by the GCF.

$$\frac{3}{3} = 1, \ \frac{15}{3} = 5$$

$$\frac{1}{5}$$

$\frac{3}{5}$ multiplied by $\frac{1}{3}$ is $\frac{1}{5}$.

Practice Problem Multiply $\frac{3}{14}$ by $\frac{5}{16}$.

Find a Reciprocal Two numbers whose product is 1 are called multiplicative inverses, or reciprocals.

Example Find the reciprocal of $\frac{3}{8}$.

Step 1 Inverse the fraction by putting the denominator on top and the numerator on the bottom.

$$\frac{8}{3}$$

The reciprocal of $\frac{3}{8}$ is $\frac{8}{3}$.

Practice Problem Find the reciprocal of $\frac{4}{9}$.

Divide Fractions To divide one fraction by another fraction, multiply the dividend by the reciprocal of the divisor. Find the simplest form if necessary.

Example 1 Divide $\frac{1}{9}$ by $\frac{1}{3}$.

Step 1 Find the reciprocal of the divisor.

The reciprocal of $\frac{1}{3}$ is $\frac{3}{1}$.

Step 2 Multiply the dividend by the reciprocal of the divisor.

$$\frac{\frac{1}{9}}{\frac{1}{3}} = \frac{1}{9} \times \frac{3}{1} = \frac{(1 \times 3)}{(9 \times 1)} = \frac{3}{9}$$

Step 3 Find the GCF.

$$\frac{3}{9} \quad (GCF, 3)$$

Step 4 Divide the numerator and denominator by the GCF.

$$\frac{3}{3} = 1, \ \frac{9}{3} = 3$$

$$\frac{1}{3}$$

$\frac{1}{9}$ divided by $\frac{1}{3}$ is $\frac{1}{3}$.

Example 2 Divide $\frac{3}{5}$ by $\frac{1}{4}$.

Step 1 Find the reciprocal of the divisor.

The reciprocal of $\frac{1}{4}$ is $\frac{4}{1}$.

Step 2 Multiply the dividend by the reciprocal of the divisor.

$$\frac{\frac{3}{5}}{\frac{1}{4}} = \frac{3}{5} \times \frac{4}{1} = \frac{(3 \times 4)}{(5 \times 1)} = \frac{12}{5}$$

$\frac{3}{5}$ divided by $\frac{1}{4}$ is $\frac{12}{5}$ or $2\frac{2}{5}$.

Practice Problem Divide $\frac{3}{11}$ by $\frac{7}{10}$.

Use Ratios

When you compare two numbers by division, you are using a ratio. Ratios can be written 3 to 5, 3:5, or $\frac{3}{5}$. Ratios, like fractions, also can be written in simplest form.

Ratios can represent probabilities, also called odds. This is a ratio that compares the number of ways a certain outcome occurs to the number of outcomes. For example, if you flip a coin 100 times, what are the odds that it will come up heads? There are two possible outcomes, heads or tails, so the odds of coming up heads are 50:100. Another way to say this is that 50 out of 100 times the coin will come up heads. In its simplest form, the ratio is 1:2.

Example 1 A chemical solution contains 40 g of salt and 64 g of baking soda. What is the ratio of salt to baking soda as a fraction in simplest form?

Step 1 Write the ratio as a fraction.
$$\frac{salt}{baking\ soda} = \frac{40}{64}$$

Step 2 Express the fraction in simplest form.
The GCF of 40 and 64 is 8.
$$\frac{40}{64} = \frac{40 \div 8}{64 \div 8} = \frac{5}{8}$$

The ratio of salt to baking soda in the sample is 5:8.

Example 2 Sean rolls a 6-sided die 6 times. What are the odds that the side with a 3 will show?

Step 1 Write the ratio as a fraction.
$$\frac{number\ of\ sides\ with\ a\ 3}{number\ of\ sides} = \frac{1}{6}$$

Step 2 Multiply by the number of attempts.
$$\frac{1}{6} \times 6\ attempts = \frac{6}{6}\ attempts = 1\ attempt$$

1 attempt out of 6 will show a 3.

Practice Problem Two metal rods measure 100 cm and 144 cm in length. What is the ratio of their lengths in simplest form?

Use Decimals

A fraction with a denominator that is a power of ten can be written as a decimal. For example, 0.27 means $\frac{27}{100}$. The decimal point separates the ones place from the tenths place.

Any fraction can be written as a decimal using division. For example, the fraction $\frac{5}{8}$ can be written as a decimal by dividing 5 by 8. Written as a decimal, it is 0.625.

Add or Subtract Decimals When adding and subtracting decimals, line up the decimal points before carrying out the operation.

Example 1 Find the sum of 47.68 and 7.80.

Step 1 Line up the decimal places when you write the numbers.

$$\begin{array}{r} 47.68 \\ + \ 7.80 \\ \hline \end{array}$$

Step 2 Add the decimals.

$$\begin{array}{r} 47.68 \\ + \ 7.80 \\ \hline 55.48 \end{array}$$

The sum of 47.68 and 7.80 is 55.48.

Example 2 Find the difference of 42.17 and 15.85.

Step 1 Line up the decimal places when you write the number.

$$\begin{array}{r} 42.17 \\ - 15.85 \\ \hline \end{array}$$

Step 2 Subtract the decimals.

$$\begin{array}{r} 42.17 \\ - 15.85 \\ \hline 26.32 \end{array}$$

The difference of 42.17 and 15.85 is 26.32.

Practice Problem Find the sum of 1.245 and 3.842.

Multiply Decimals To multiply decimals, multiply the numbers like any other number, ignoring the decimal point. Count the decimal places in each factor. The product will have the same number of decimal places as the sum of the decimal places in the factors.

Example Multiply 2.4 by 5.9.

Step 1 Multiply the factors like two whole numbers.

$24 \times 59 = 1416$

Step 2 Find the sum of the number of decimal places in the factors. Each factor has one decimal place, for a sum of two decimal places.

Step 3 The product will have two decimal places.

14.16

The product of 2.4 and 5.9 is 14.16.

Practice Problem Multiply 4.6 by 2.2.

Divide Decimals When dividing decimals, change the divisor to a whole number. To do this, multiply both the divisor and the dividend by the same power of ten. Then place the decimal point in the quotient directly above the decimal point in the dividend. Then divide as you do with whole numbers.

Example Divide 8.84 by 3.4.

Step 1 Multiply both factors by 10.

$3.4 \times 10 = 34, 8.84 \times 10 = 88.4$

Step 2 Divide 88.4 by 34.

$$
\begin{array}{r}
2.6 \\
34\overline{)88.4} \\
-\underline{68} \\
204 \\
-\underline{204} \\
0
\end{array}
$$

8.84 divided by 3.4 is 2.6.

Practice Problem Divide 75.6 by 3.6.

Use Proportions

An equation that shows that two ratios are equivalent is a proportion. The ratios $\frac{2}{4}$ and $\frac{5}{10}$ are equivalent, so they can be written as $\frac{2}{4} = \frac{5}{10}$. This equation is a proportion.

When two ratios form a proportion, the cross products are equal. To find the cross products in the proportion $\frac{2}{4} = \frac{5}{10}$, multiply the 2 and the 10, and the 4 and the 5. Therefore $2 \times 10 = 4 \times 5$, or $20 = 20$.

Because you know that both proportions are equal, you can use cross products to find a missing term in a proportion. This is known as solving the proportion.

Example The heights of a tree and a pole are proportional to the lengths of their shadows. The tree casts a shadow of 24 m when a 6-m pole casts a shadow of 4 m. What is the height of the tree?

Step 1 Write a proportion.

$$\frac{\text{height of tree}}{\text{height of pole}} = \frac{\text{length of tree's shadow}}{\text{length of pole's shadow}}$$

Step 2 Substitute the known values into the proportion. Let h represent the unknown value, the height of the tree.

$$\frac{h}{6} = \frac{24}{4}$$

Step 3 Find the cross products.

$h \times 4 = 6 \times 24$

Step 4 Simplify the equation.

$4h = 144$

Step 5 Divide each side by 4.

$$\frac{4h}{4} = \frac{144}{4}$$
$$h = 36$$

The height of the tree is 36 m.

Practice Problem The ratios of the weights of two objects on the Moon and on Earth are in proportion. A rock weighing 3 N on the Moon weighs 18 N on Earth. How much would a rock that weighs 5 N on the Moon weigh on Earth?

Use Percentages

The word *percent* means "out of one hundred." It is a ratio that compares a number to 100. Suppose you read that 77 percent of the Earth's surface is covered by water. That is the same as reading that the fraction of the Earth's surface covered by water is $\frac{77}{100}$. To express a fraction as a percent, first find the equivalent decimal for the fraction. Then, multiply the decimal by 100 and add the percent symbol.

Example Express $\frac{13}{20}$ as a percent.

Step 1 Find the equivalent decimal for the fraction.

$$\begin{array}{r} 0.65 \\ 20)\overline{13.00} \\ \underline{12\,0} \\ 1\,00 \\ \underline{1\,00} \\ 0 \end{array}$$

Step 2 Rewrite the fraction $\frac{13}{20}$ as 0.65.

Step 3 Multiply 0.65 by 100 and add the % sign.
$$0.65 \times 100 = 65 = 65\%$$

So, $\frac{13}{20} = 65\%$.

This also can be solved as a proportion.

Example Express $\frac{13}{20}$ as a percent.

Step 1 Write a proportion.
$$\frac{13}{20} = \frac{x}{100}$$

Step 2 Find the cross products.
$$1300 = 20x$$

Step 3 Divide each side by 20.
$$\frac{1300}{20} = \frac{20x}{20}$$
$$65\% = x$$

Practice Problem In one year, 73 of 365 days were rainy in one city. What percent of the days in that city were rainy?

Solve One-Step Equations

A statement that two things are equal is an equation. For example, $A = B$ is an equation that states that A is equal to B.

An equation is solved when a variable is replaced with a value that makes both sides of the equation equal. To make both sides equal the inverse operation is used. Addition and subtraction are inverses, and multiplication and division are inverses.

Example 1 Solve the equation $x - 10 = 35$.

Step 1 Find the solution by adding 10 to each side of the equation.
$$x - 10 = 35$$
$$x - 10 + 10 = 35 + 10$$
$$x = 45$$

Step 2 Check the solution.
$$x - 10 = 35$$
$$45 - 10 = 35$$
$$35 = 35$$

Both sides of the equation are equal, so $x = 45$.

Example 2 In the formula $a = bc$, find the value of c if $a = 20$ and $b = 2$.

Step 1 Rearrange the formula so the unknown value is by itself on one side of the equation by dividing both sides by b.
$$a = bc$$
$$\frac{a}{b} = \frac{bc}{b}$$
$$\frac{a}{b} = c$$

Step 2 Replace the variables a and b with the values that are given.
$$\frac{a}{b} = c$$
$$\frac{20}{2} = c$$
$$10 = c$$

Step 3 Check the solution.
$$a = bc$$
$$20 = 2 \times 10$$
$$20 = 20$$

Both sides of the equation are equal, so $c = 10$ is the solution when $a = 20$ and $b = 2$.

Practice Problem In the formula $h = gd$, find the value of d if $g = 12.3$ and $h = 17.4$.

Use Statistics

The branch of mathematics that deals with collecting, analyzing, and presenting data is statistics. In statistics, there are three common ways to summarize data with a single number—the mean, the median, and the mode.

The **mean** of a set of data is the arithmetic average. It is found by adding the numbers in the data set and dividing by the number of items in the set.

The **median** is the middle number in a set of data when the data are arranged in numerical order. If there were an even number of data points, the median would be the mean of the two middle numbers.

The **mode** of a set of data is the number or item that appears most often.

Another number that often is used to describe a set of data is the range. The **range** is the difference between the largest number and the smallest number in a set of data.

A **frequency table** shows how many times each piece of data occurs, usually in a survey. **Table 2** below shows the results of a student survey on favorite color.

Table 2 Student Color Choice		
Color	**Tally**	**Frequency**
red	||||	4
blue	卅	5
black	||	2
green	|||	3
purple	卅 ||	7
yellow	卅 |	6

Based on the frequency table data, which color is the favorite?

Example The speeds (in m/s) for a race car during five different time trials are 39, 37, 44, 36, and 44.

To find the mean:

Step 1 Find the sum of the numbers.

$$39 + 37 + 44 + 36 + 44 = 200$$

Step 2 Divide the sum by the number of items, which is 5.

$$200 \div 5 = 40$$

The mean is 40 m/s.

To find the median:

Step 1 Arrange the measures from least to greatest.

36, 37, 39, 44, 44

Step 2 Determine the middle measure.

36, 37, <u>39</u>, 44, 44

The median is 39 m/s.

To find the mode:

Step 1 Group the numbers that are the same together.

44, 44, 36, 37, 39

Step 2 Determine the number that occurs most in the set.

<u>44, 44</u>, 36, 37, 39

The mode is 44 m/s.

To find the range:

Step 1 Arrange the measures from largest to smallest.

44, 44, 39, 37, 36

Step 2 Determine the largest and smallest measures in the set.

<u>44</u>, 44, 39, 37, <u>36</u>

Step 3 Find the difference between the largest and smallest measures.

$$44 - 36 = 8$$

The range is 8 m/s.

Practice Problem Find the mean, median, mode, and range for the data set 8, 4, 12, 8, 11, 14, 16.

Use Geometry

The branch of mathematics that deals with the measurement, properties, and relationships of points, lines, angles, surfaces, and solids is called geometry.

Perimeter The **perimeter** (P) is the distance around a geometric figure. To find the perimeter of a rectangle, add the length and width and multiply that sum by two, or $2(l + w)$. To find perimeters of irregular figures, add the length of the sides.

Example 1 Find the perimeter of a rectangle that is 3 m long and 5 m wide.

Step 1 You know that the perimeter is 2 times the sum of the width and length.
$P = 2(3\text{ m} + 5\text{ m})$

Step 2 Find the sum of the width and length.
$P = 2(8\text{ m})$

Step 3 Multiply by 2.
$P = 16\text{ m}$

The perimeter is 16 m.

Example 2 Find the perimeter of a shape with sides measuring 2 cm, 5 cm, 6 cm, 3 cm.

Step 1 You know that the perimeter is the sum of all the sides.
$P = 2 + 5 + 6 + 3$

Step 2 Find the sum of the sides.
$P = 2 + 5 + 6 + 3$
$P = 16$

The perimeter is 16 cm.

Practice Problem Find the perimeter of a rectangle with a length of 18 m and a width of 7 m.

Practice Problem Find the perimeter of a triangle measuring 1.6 cm by 2.4 cm by 2.4 cm.

Area of a Rectangle The **area** (A) is the number of square units needed to cover a surface. To find the area of a rectangle, multiply the length times the width, or $l \times w$. When finding area, the units also are multiplied. Area is given in square units.

Example Find the area of a rectangle with a length of 1 cm and a width of 10 cm.

Step 1 You know that the area is the length multiplied by the width.
$A = (1\text{ cm} \times 10\text{ cm})$

Step 2 Multiply the length by the width. Also multiply the units.
$A = 10\text{ cm}^2$

The area is 10 cm^2.

Practice Problem Find the area of a square whose sides measure 4 m.

Area of a Triangle To find the area of a triangle, use the formula:

$$A = \frac{1}{2}(\text{base} \times \text{height})$$

The base of a triangle can be any of its sides. The height is the perpendicular distance from a base to the opposite endpoint, or vertex.

Example Find the area of a triangle with a base of 18 m and a height of 7 m.

Step 1 You know that the area is $\frac{1}{2}$ the base times the height.
$A = \frac{1}{2}(18\text{ m} \times 7\text{ m})$

Step 2 Multiply $\frac{1}{2}$ by the product of 18×7. Multiply the units.
$A = \frac{1}{2}(126\text{ m}^2)$
$A = 63\text{ m}^2$

The area is 63 m^2.

Practice Problem Find the area of a triangle with a base of 27 cm and a height of 17 cm.

Circumference of a Circle The **diameter** (d) of a circle is the distance across the circle through its center, and the **radius** (r) is the distance from the center to any point on the circle. The radius is half of the diameter. The distance around the circle is called the **circumference** (C). The formula for finding the circumference is:

$$C = 2\pi r \ \ or \ \ C = \pi d$$

The circumference divided by the diameter is always equal to 3.1415926... This nonterminating and nonrepeating number is represented by the Greek letter π (pi). An approximation often used for π is 3.14.

Example 1 Find the circumference of a circle with a radius of 3 m.

Step 1 You know the formula for the circumference is 2 times the radius times π.
$C = 2\pi(3)$

Step 2 Multiply 2 times the radius.
$C = 6\pi$

Step 3 Multiply by π.
$C = 19$ m

The circumference is 19 m.

Example 2 Find the circumference of a circle with a diameter of 24.0 cm.

Step 1 You know the formula for the circumference is the diameter times π.
$C = \pi(24.0)$

Step 2 Multiply the diameter by π.
$C = 75.4$ cm

The circumference is 75.4 cm.

Practice Problem Find the circumference of a circle with a radius of 19 cm.

Area of a Circle The formula for the area of a circle is:
$A = \pi r^2$

Example 1 Find the area of a circle with a radius of 4.0 cm.

Step 1 $A = \pi(4.0)^2$

Step 2 Find the square of the radius.
$A = 16\pi$

Step 3 Multiply the square of the radius by π.
$A = 50$ cm^2

The area of the circle is 50 cm^2.

Example 2 Find the area of a circle with a radius of 225 m.

Step 1 $A = \pi(225)^2$

Step 2 Find the square of the radius.
$A = 50625\pi$

Step 3 Multiply the square of the radius by π.
$A = 158962.5$

The area of the circle is 158,962 m^2.

Example 3 Find the area of a circle whose diameter is 20.0 mm.

Step 1 You know the formula for the area of a circle is the square of the radius times π, and that the radius is half of the diameter.
$A = \pi\left(\dfrac{20.0}{2}\right)^2$

Step 2 Find the radius.
$A = \pi(10.0)^2$

Step 3 Find the square of the radius.
$A = 100\pi$

Step 4 Multiply the square of the radius by π.
$A = 314$ mm^2

The area is 314 mm^2.

Practice Problem Find the area of a circle with a radius of 16 m.

Volume The measure of space occupied by a solid is the **volume** (V). To find the volume of a rectangular solid multiply the length times width times height, or $V = l \times w \times h$. It is measured in cubic units, such as cubic centimeters (cm^3).

Example Find the volume of a rectangular solid with a length of 2.0 m, a width of 4.0 m, and a height of 3.0 m.

Step 1 You know the formula for volume is the length times the width times the height.
$$V = 2.0 \text{ m} \times 4.0 \text{ m} \times 3.0 \text{ m}$$

Step 2 Multiply the length times the width times the height.
$$V = 24 \text{ m}^3$$

The volume is 24 m³.

Practice Problem Find the volume of a rectangular solid that is 8 m long, 4 m wide, and 4 m high.

To find the volume of other solids, multiply the area of the base times the height.

Example 1 Find the volume of a solid that has a triangular base with a length of 8.0 m and a height of 7.0 m. The height of the entire solid is 15.0 m.

Step 1 You know that the base is a triangle, and the area of a triangle is $\frac{1}{2}$ the base times the height, and the volume is the area of the base times the height.
$$V = \left[\frac{1}{2}(b \times h)\right] \times 15$$

Step 2 Find the area of the base.
$$V = \left[\frac{1}{2}(8 \times 7)\right] \times 15$$
$$V = \left(\frac{1}{2} \times 56\right) \times 15$$

Step 3 Multiply the area of the base by the height of the solid.
$$V = 28 \times 15$$
$$V = 420 \text{ m}^3$$

The volume is 420 m³.

Example 2 Find the volume of a cylinder that has a base with a radius of 12.0 cm, and a height of 21.0 cm.

Step 1 You know that the base is a circle, and the area of a circle is the square of the radius times π, and the volume is the area of the base times the height.
$$V = (\pi r^2) \times 21$$
$$V = (\pi 12^2) \times 21$$

Step 2 Find the area of the base.
$$V = 144\pi \times 21$$
$$V = 452 \times 21$$

Step 3 Multiply the area of the base by the height of the solid.
$$V = 9490 \text{ cm}^3$$

The volume is 9490 cm³.

Example 3 Find the volume of a cylinder that has a diameter of 15 mm and a height of 4.8 mm.

Step 1 You know that the base is a circle with an area equal to the square of the radius times π. The radius is one-half the diameter. The volume is the area of the base times the height.
$$V = (\pi r^2) \times 4.8$$
$$V = \left[\pi\left(\frac{1}{2} \times 15\right)^2\right] \times 4.8$$
$$V = (\pi 7.5^2) \times 4.8$$

Step 2 Find the area of the base.
$$V = 56.25\pi \times 4.8$$
$$V = 176.63 \times 4.8$$

Step 3 Multiply the area of the base by the height of the solid.
$$V = 847.8$$

The volume is 847.8 mm³.

Practice Problem Find the volume of a cylinder with a diameter of 7 cm in the base and a height of 16 cm.

Science Applications

Measure in SI

The metric system of measurement was developed in 1795. A modern form of the metric system, called the International System (SI), was adopted in 1960 and provides the standard measurements that all scientists around the world can understand.

The SI system is convenient because unit sizes vary by powers of 10. Prefixes are used to name units. Look at **Table 3** for some common SI prefixes and their meanings.

Table 3 Common SI Prefixes			
Prefix	**Symbol**	**Meaning**	
kilo-	k	1,000	thousand
hecto-	h	100	hundred
deka-	da	10	ten
deci-	d	0.1	tenth
centi-	c	0.01	hundredth
milli-	m	0.001	thousandth

Example How many grams equal one kilogram?

Step 1 Find the prefix *kilo* in **Table 3**.

Step 2 Using **Table 3,** determine the meaning of *kilo.* According to the table, it means 1,000. When the prefix *kilo* is added to a unit, it means that there are 1,000 of the units in a "*kilo*unit."

Step 3 Apply the prefix to the units in the question. The units in the question are grams. There are 1,000 grams in a kilogram.

Practice Problem Is a milligram larger or smaller than a gram? How many of the smaller units equal one larger unit? What fraction of the larger unit does one smaller unit represent?

Dimensional Analysis

Convert SI Units In science, quantities such as length, mass, and time sometimes are measured using different units. A process called dimensional analysis can be used to change one unit of measure to another. This process involves multiplying your starting quantity and units by one or more conversion factors. A conversion factor is a ratio equal to one and can be made from any two equal quantities with different units. If 1,000 mL equal 1 L then two ratios can be made.

$$\frac{1{,}000 \text{ mL}}{1 \text{ L}} = \frac{1 \text{ L}}{1{,}000 \text{ mL}} = 1$$

One can covert between units in the SI system by using the equivalents in **Table 3** to make conversion factors.

Example 1 How many cm are in 4 m?

Step 1 Write conversion factors for the units given. From **Table 3,** you know that 100 cm = 1 m. The conversion factors are

$$\frac{100 \text{ cm}}{1 \text{ m}} \quad and \quad \frac{1 \text{ m}}{100 \text{ cm}}$$

Step 2 Decide which conversion factor to use. Select the factor that has the units you are converting from (m) in the denominator and the units you are converting to (cm) in the numerator.

$$\frac{100 \text{ cm}}{1 \text{ m}}$$

Step 3 Multiply the starting quantity and units by the conversion factor. Cancel the starting units with the units in the denominator. There are 400 cm in 4 m.

$$4 \text{ m} \times \frac{100 \text{ cm}}{1 \text{ m}} = 400 \text{ cm}$$

Practice Problem How many milligrams are in one kilogram? (Hint: You will need to use two conversion factors from **Table 3**.)

Table 4 Unit System Equivalents

Type of Measurement	Equivalent
Length	1 in = 2.54 cm
	1 yd = 0.91 m
	1 mi = 1.61 km
Mass and Weight*	1 oz = 28.35 g
	1 lb = 0.45 kg
	1 ton (short) = 0.91 tonnes (metric tons)
	1 lb = 4.45 N
Volume	$1\ in^3 = 16.39\ cm^3$
	1 qt = 0.95 L
	1 gal = 3.78 L
Area	$1\ in^2 = 6.45\ cm^2$
	$1\ yd^2 = 0.83\ m^2$
	$1\ mi^2 = 2.59\ km^2$
	1 acre = 0.40 hectares
Temperature	$°C = \dfrac{(°F - 32)}{1.8}$
	$K = °C + 273$

*Weight is measured in standard Earth gravity.

Convert Between Unit Systems **Table 4** gives a list of equivalents that can be used to convert between English and SI units.

Example If a meterstick has a length of 100 cm, how long is the meterstick in inches?

Step 1 Write the conversion factors for the units given. From **Table 4,** 1 in = 2.54 cm.

$$\frac{1\ in}{2.54\ cm}\quad and\quad \frac{2.54\ cm}{1\ in}$$

Step 2 Determine which conversion factor to use. You are converting from cm to in. Use the conversion factor with cm on the bottom.

$$\frac{1\ in}{2.54\ cm}$$

Step 3 Multiply the starting quantity and units by the conversion factor. Cancel the starting units with the units in the denominator. Round your answer based on the number of significant figures in the conversion factor.

$$100\ \cancel{cm} \times \frac{1\ in}{2.54\ \cancel{cm}} = 39.37\ in$$

The meterstick is 39.4 in long.

Practice Problem A book has a mass of 5 lbs. What is the mass of the book in kg?

Practice Problem Use the equivalent for in and cm (1 in = 2.54 cm) to show how $1\ in^3 = 16.39\ cm^3$.

Precision and Significant Digits

When you make a measurement, the value you record depends on the precision of the measuring instrument. This precision is represented by the number of significant digits recorded in the measurement. When counting the number of significant digits, all digits are counted except zeros at the end of a number with no decimal point such as 2,050, and zeros at the beginning of a decimal such as 0.03020. When adding or subtracting numbers with different precision, round the answer to the smallest number of decimal places of any number in the sum or difference. When multiplying or dividing, the answer is rounded to the smallest number of significant digits of any number being multiplied or divided.

Example The lengths 5.28 and 5.2 are measured in meters. Find the sum of these lengths and record your answer using the correct number of significant digits.

Step 1 Find the sum.

5.28 m	2 digits after the decimal
+ 5.2 m	1 digit after the decimal
10.48 m	

Step 2 Round to one digit after the decimal because the least number of digits after the decimal of the numbers being added is 1.

The sum is 10.5 m.

Practice Problem How many significant digits are in the measurement 7,071,301 m? How many significant digits are in the measurement 0.003010 g?

Practice Problem Multiply 5.28 and 5.2 using the rule for multiplying and dividing. Record the answer using the correct number of significant digits.

Scientific Notation

Many times numbers used in science are very small or very large. Because these numbers are difficult to work with scientists use scientific notation. To write numbers in scientific notation, move the decimal point until only one non-zero digit remains on the left. Then count the number of places you moved the decimal point and use that number as a power of ten. For example, the average distance from the Sun to Mars is 227,800,000,000 m. In scientific notation, this distance is 2.278×10^{11} m. Because you moved the decimal point to the left, the number is a positive power of ten.

The mass of an electron is about 0.000 000 000 000 000 000 000 000 000 000 911 kg. Expressed in scientific notation, this mass is 9.11×10^{-31} kg. Because the decimal point was moved to the right, the number is a negative power of ten.

Example Earth is 149,600,000 km from the Sun. Express this in scientific notation.

Step 1 Move the decimal point until one non-zero digit remains on the left.
1.496 000 00

Step 2 Count the number of decimal places you have moved. In this case, eight.

Step 3 Show that number as a power of ten, 10^8.

The Earth is 1.496×10^8 km from the Sun.

Practice Problem How many significant digits are in 149,600,000 km? How many significant digits are in 1.496×10^8 km?

Practice Problem Parts used in a high performance car must be measured to 7×10^{-6} m. Express this number as a decimal.

Practice Problem A CD is spinning at 539 revolutions per minute. Express this number in scientific notation.

Math Skill Handbook

Make and Use Graphs

Data in tables can be displayed in a graph—a visual representation of data. Common graph types include line graphs, bar graphs, and circle graphs.

Line Graph A line graph shows a relationship between two variables that change continuously. The independent variable is changed and is plotted on the *x*-axis. The dependent variable is observed, and is plotted on the *y*-axis.

Example Draw a line graph of the data below from a cyclist in a long-distance race.

Table 5 Bicycle Race Data

Time (h)	Distance (km)
0	0
1	8
2	16
3	24
4	32
5	40

Step 1 Determine the *x*-axis and *y*-axis variables. Time varies independently of distance and is plotted on the *x*-axis. Distance is dependent on time and is plotted on the *y*-axis.

Step 2 Determine the scale of each axis. The *x*-axis data ranges from 0 to 5. The *y*-axis data ranges from 0 to 40.

Step 3 Using graph paper, draw and label the axes. Include units in the labels.

Step 4 Draw a point at the intersection of the time value on the *x*-axis and corresponding distance value on the *y*-axis. Connect the points and label the graph with a title, as shown in **Figure 20.**

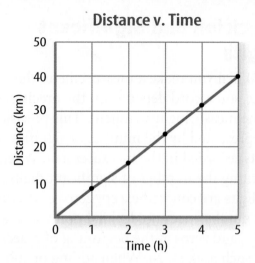

Distance v. Time

Figure 20 This line graph shows the relationship between distance and time during a bicycle ride.

Practice Problem A puppy's shoulder height is measured during the first year of her life. The following measurements were collected: (3 mo, 52 cm), (6 mo, 72 cm), (9 mo, 83 cm), (12 mo, 86 cm). Graph this data.

Find a Slope The slope of a straight line is the ratio of the vertical change, rise, to the horizontal change, run.

$$\text{Slope} = \frac{\text{vertical change (rise)}}{\text{horizontal change (run)}} = \frac{\text{change in } y}{\text{change in } x}$$

Example Find the slope of the graph in **Figure 20.**

Step 1 You know that the slope is the change in *y* divided by the change in *x*.
$$\text{Slope} = \frac{\text{change in } y}{\text{change in } x}$$

Step 2 Determine the data points you will be using. For a straight line, choose the two sets of points that are the farthest apart.
$$\text{Slope} = \frac{(40-0)\text{ km}}{(5-0)\text{ hr}}$$

Step 3 Find the change in *y* and *x*.
$$\text{Slope} = \frac{40\text{ km}}{5\text{h}}$$

Step 4 Divide the change in *y* by the change in *x*.
$$\text{Slope} = \frac{8\text{ km}}{\text{h}}$$

The slope of the graph is 8 km/h.

Bar Graph To compare data that does not change continuously you might choose a bar graph. A bar graph uses bars to show the relationships between variables. The *x*-axis variable is divided into parts. The parts can be numbers such as years, or a category such as a type of animal. The *y*-axis is a number and increases continuously along the axis.

Example A recycling center collects 4.0 kg of aluminum on Monday, 1.0 kg on Wednesday, and 2.0 kg on Friday. Create a bar graph of this data.

Step 1 Select the *x*-axis and *y*-axis variables. The measured numbers (the masses of aluminum) should be placed on the *y*-axis. The variable divided into parts (collection days) is placed on the *x*-axis.

Step 2 Create a graph grid like you would for a line graph. Include labels and units.

Step 3 For each measured number, draw a vertical bar above the *x*-axis value up to the *y*-axis value. For the first data point, draw a vertical bar above Monday up to 4.0 kg.

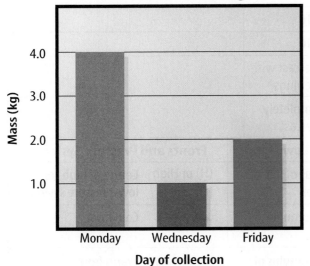

Aluminum Collected During Week

Practice Problem Draw a bar graph of the gases in air: 78% nitrogen, 21% oxygen, 1% other gases.

Circle Graph To display data as parts of a whole, you might use a circle graph. A circle graph is a circle divided into sections that represent the relative size of each piece of data. The entire circle represents 100%, half represents 50%, and so on.

Example Air is made up of 78% nitrogen, 21% oxygen, and 1% other gases. Display the composition of air in a circle graph.

Step 1 Multiply each percent by 360° and divide by 100 to find the angle of each section in the circle.

$$78\% \times \frac{360°}{100} = 280.8°$$

$$21\% \times \frac{360°}{100} = 75.6°$$

$$1\% \times \frac{360°}{100} = 3.6°$$

Step 2 Use a compass to draw a circle and to mark the center of the circle. Draw a straight line from the center to the edge of the circle.

Step 3 Use a protractor and the angles you calculated to divide the circle into parts. Place the center of the protractor over the center of the circle and line the base of the protractor over the straight line.

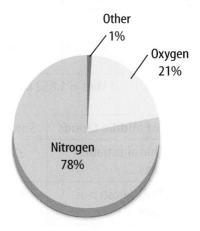

Practice Problem Draw a circle graph to represent the amount of aluminum collected during the week shown in the bar graph to the left.

Weather Map Symbols

Sample Station Model

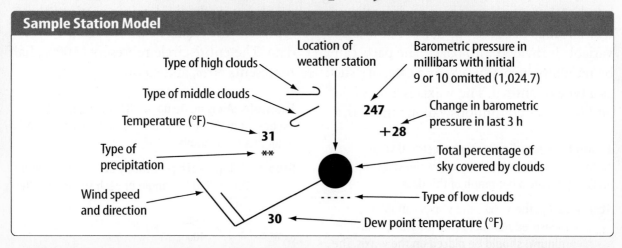

- Location of weather station
- Type of high clouds
- Type of middle clouds
- Temperature (°F) — **31**
- Type of precipitation — ✱✱
- Wind speed and direction
- Barometric pressure in millibars with initial 9 or 10 omitted (1,024.7) — **247**
- Change in barometric pressure in last 3 h — **+28**
- Total percentage of sky covered by clouds
- Type of low clouds — - - - - -
- Dew point temperature (°F) — **30**

Sample Plotted Report at Each Station

Precipitation		Wind Speed and Direction		Sky Coverage		Some Types of High Clouds	
≡	Fog	◯	0 calm	◯	No cover	⌐	Scattered cirrus
★	Snow	╱	1–2 knots	◔	1/10 or less	⌐⌐	Dense cirrus in patches
●	Rain	⌐	3–7 knots	◔	2/10 to 3/10	⌐	Veil of cirrus covering entire sky
⊥	Thunderstorm	⌐	8–12 knots	◑	4/10	⌐	Cirrus not covering entire sky
,	Drizzle	⫳	13–17 knots	◐	–		
▽	Showers	⫳	18–22 knots	◕	6/10		
		⫶⫳	23–27 knots	◕	7/10		
		⫰	48–52 knots	◉	Overcast with openings		
		1 knot = 1.852 km/h		●	Completely overcast		

Some Types of Middle Clouds		Some Types of Low Clouds		Fronts and Pressure Systems	
╱	Thin altostratus layer	⌒	Cumulus of fair weather	(H) or High (L) or Low	Center of high- or low-pressure system
╱╱	Thick altostratus layer	⌣	Stratocumulus	▲▲▲▲	Cold front
╱	Thin altostratus in patches	- - - - -	Fractocumulus of bad weather	●●●●	Warm front
╱	Thin altostratus in bands	—	Stratus of fair weather	▲●▲●	Occluded front
				●▲●▲	Stationary front

Rocks

Rocks		
Rock Type	**Rock Name**	**Characteristics**
Igneous (intrusive)	Granite	Large mineral grains of quartz, feldspar, hornblende, and mica. Usually light in color.
	Diorite	Large mineral grains of feldspar, hornblende, and mica. Less quartz than granite. Intermediate in color.
	Gabbro	Large mineral grains of feldspar, augite, and olivine. No quartz. Dark in color.
Igneous (extrusive)	Rhyolite	Small mineral grains of quartz, feldspar, hornblende, and mica, or no visible grains. Light in color.
	Andesite	Small mineral grains of feldspar, hornblende, and mica or no visible grains. Intermediate in color.
	Basalt	Small mineral grains of feldspar, augite, and possibly olivine or no visible grains. No quartz. Dark in color.
	Obsidian	Glassy texture. No visible grains. Volcanic glass. Fracture looks like broken glass.
	Pumice	Frothy texture. Floats in water. Usually light in color.
Sedimentary (detrital)	Conglomerate	Coarse grained. Gravel or pebble-size grains.
	Sandstone	Sand-sized grains 1/16 to 2 mm.
	Siltstone	Grains are smaller than sand but larger than clay.
	Shale	Smallest grains. Often dark in color. Usually platy.
Sedimentary (chemical or organic)	Limestone	Major mineral is calcite. Usually forms in oceans and lakes. Often contains fossils.
	Coal	Forms in swampy areas. Compacted layers of organic material, mainly plant remains.
Sedimentary (chemical)	Rock Salt	Commonly forms by the evaporation of seawater.
Metamorphic (foliated)	Gneiss	Banding due to alternate layers of different minerals, of different colors. Parent rock often is granite.
	Schist	Parallel arrangement of sheetlike minerals, mainly micas. Forms from different parent rocks.
	Phyllite	Shiny or silky appearance. May look wrinkled. Common parent rocks are shale and slate.
	Slate	Harder, denser, and shinier than shale. Common parent rock is shale.
Metamorphic (nonfoliated)	Marble	Calcite or dolomite. Common parent rock is limestone.
	Soapstone	Mainly of talc. Soft with greasy feel.
	Quartzite	Hard with interlocking quartz crystals. Common parent rock is sandstone.

Minerals

Minerals					
Mineral (formula)	**Color**	**Streak**	**Hardness**	**Breakage Pattern**	**Uses and Other Properties**
Graphite (C)	black to gray	black to gray	1–1.5	basal cleavage (scales)	pencil lead, lubricants for locks, rods to control some small nuclear reactions, battery poles
Galena (PbS)	gray	gray to black	2.5	cubic cleavage perfect	source of lead, used for pipes, shields for X rays, fishing equipment sinkers
Hematite (Fe_2O_3)	black or reddish-brown	reddish-brown	5.5–6.5	irregular fracture	source of iron; converted to pig iron, made into steel
Magnetite (Fe_3O_4)	black	black	6	conchoidal fracture	source of iron, attracts a magnet
Pyrite (FeS_2)	light, brassy, yellow	greenish-black	6–6.5	uneven fracture	fool's gold
Talc ($Mg_3 Si_4O_{10} (OH)_2$)	white, greenish	white	1	cleavage in one direction	used for talcum powder, sculptures, paper, and tabletops
Gypsum ($CaSO_4 \cdot 2H_2O$)	colorless, gray, white, brown	white	2	basal cleavage	used in plaster of paris and dry wall for building construction
Sphalerite (ZnS)	brown, reddish-brown, greenish	light to dark brown	3.5–4	cleavage in six directions	main ore of zinc; used in paints, dyes, and medicine
Muscovite ($KAl_3Si_3 O_{10}(OH)_2$)	white, light gray, yellow, rose, green	colorless	2–2.5	basal cleavage	occurs in large, flexible plates; used as an insulator in electrical equipment, lubricant
Biotite ($K(Mg,Fe)_3 (AlSi_3O_{10}) (OH)_2$)	black to dark brown	colorless	2.5–3	basal cleavage	occurs in large, flexible plates
Halite (NaCl)	colorless, red, white, blue	colorless	2.5	cubic cleavage	salt; soluble in water; a preservative

Minerals

Minerals					
Mineral (formula)	**Color**	**Streak**	**Hardness**	**Breakage Pattern**	**Uses and Other Properties**
Calcite ($CaCO_3$)	colorless, white, pale blue	colorless, white	3	cleavage in three directions	fizzes when HCl is added; used in cements and other building materials
Dolomite ($CaMg(CO_3)_2$)	colorless, white, pink, green, gray, black	white	3.5–4	cleavage in three directions	concrete and cement; used as an ornamental building stone
Fluorite (CaF_2)	colorless, white, blue, green, red, yellow, purple	colorless	4	cleavage in four directions	used in the manufacture of optical equipment; glows under ultraviolet light
Hornblende ($(CaNa)_{2-3}$ $(Mg,Al,$ $Fe)_5-(Al,Si)_2$ Si_6O_{22} $(OH)_2$)	green to black	gray to white	5–6	cleavage in two directions	will transmit light on thin edges; 6-sided cross section
Feldspar ($KAlSi_3O_8$) $(NaAl$ $Si_3O_8),$ $(CaAl_2Si_2$ O_8)	colorless, white to gray, green	colorless	6	two cleavage planes meet at 90° angle	used in the manufacture of ceramics
Augite ((Ca,Na) (Mg,Fe,Al) $(Al,Si)_2 O_6$)	black	colorless	6	cleavage in two directions	square or 8-sided cross section
Olivine ($(Mg,Fe)_2$ SiO_4)	olive, green	none	6.5–7	conchoidal fracture	gemstones, refractory sand
Quartz (SiO_2)	colorless, various colors	none	7	conchoidal fracture	used in glass manufacture, electronic equipment, radios, computers, watches, gemstones

PERIODIC TABLE OF THE ELEMENTS

Columns of elements are called groups. Elements in the same group have similar chemical properties.

Gas

Liquid

Solid

Synthetic

Element ——— Hydrogen
Atomic number ——— 1
Symbol ——— **H**
Atomic mass ——— 1.008

State of matter

The first three symbols tell you the state of matter of the element at room temperature. The fourth symbol identifies elements that are not present in significant amounts on Earth. Useful amounts are made synthetically.

1

	1	2							3	4	5	6	7	8	9

1
Hydrogen
1
H
1.008

2
Lithium
3
Li
6.941

Beryllium
4
Be
9.012

3
Sodium
11
Na
22.990

Magnesium
12
Mg
24.305

3 **4** **5** **6** **7** **8** **9**

4
Potassium
19
K
39.098

Calcium
20
Ca
40.078

Scandium
21
Sc
44.956

Titanium
22
Ti
47.867

Vanadium
23
V
50.942

Chromium
24
Cr
51.996

Manganese
25
Mn
54.938

Iron
26
Fe
55.845

Cobalt
27
Co
58.933

5
Rubidium
37
Rb
85.468

Strontium
38
Sr
87.62

Yttrium
39
Y
88.906

Zirconium
40
Zr
91.224

Niobium
41
Nb
92.906

Molybdenum
42
Mo
95.94

Technetium
43
Tc
(98)

Ruthenium
44
Ru
101.07

Rhodium
45
Rh
102.906

6
Cesium
55
Cs
132.905

Barium
56
Ba
137.327

Lanthanum
57
La
138.906

Hafnium
72
Hf
178.49

Tantalum
73
Ta
180.948

Tungsten
74
W
183.84

Rhenium
75
Re
186.207

Osmium
76
Os
190.23

Iridium
77
Ir
192.217

7
Francium
87
Fr
(223)

Radium
88
Ra
(226)

Actinium
89
Ac
(227)

Rutherfordium
104
Rf
(261)

Dubnium
105
Db
(262)

Seaborgium
106
Sg
(266)

Bohrium
107
Bh
(264)

Hassium
108
Hs
(277)

Meitnerium
109
Mt
(268)

The number in parentheses is the mass number of the longest-lived isotope for that element.

Rows of elements are called periods. Atomic number increases across a period.

The arrow shows where these elements would fit into the periodic table. They are moved to the bottom of the table to save space.

Lanthanide series

Cerium
58
Ce
140.116

Praseodymium
59
Pr
140.908

Neodymium
60
Nd
144.24

Promethium
61
Pm
(145)

Samarium
62
Sm
150.36

Actinide series

Thorium
90
Th
232.038

Protactinium
91
Pa
231.036

Uranium
92
U
238.029

Neptunium
93
Np
(237)

Plutonium
94
Pu
(244)

Metal

Metalloid

Nonmetal

The color of an element's block tells you if the element is a metal, nonmetal, or metalloid.

Science Online
Visit bookg.msscience.com for updates to the periodic table.

18

Helium
2
He
4.003

13 **14** **15** **16** **17**

Boron	Carbon	Nitrogen	Oxygen	Fluorine	Neon
5	6	7	8	9	10
B	**C**	**N**	**O**	**F**	**Ne**
10.811	12.011	14.007	15.999	18.998	20.180

Aluminum	Silicon	Phosphorus	Sulfur	Chlorine	Argon
13	14	15	16	17	18
Al	**Si**	**P**	**S**	**Cl**	**Ar**
26.982	28.086	30.974	32.065	35.453	39.948

10 **11** **12**

Nickel	Copper	Zinc	Gallium	Germanium	Arsenic	Selenium	Bromine	Krypton
28	29	30	31	32	33	34	35	36
Ni	**Cu**	**Zn**	**Ga**	**Ge**	**As**	**Se**	**Br**	**Kr**
58.693	63.546	65.409	69.723	72.64	74.922	78.96	79.904	83.798

Palladium	Silver	Cadmium	Indium	Tin	Antimony	Tellurium	Iodine	Xenon
46	47	48	49	50	51	52	53	54
Pd	**Ag**	**Cd**	**In**	**Sn**	**Sb**	**Te**	**I**	**Xe**
106.42	107.868	112.411	114.818	118.710	121.760	127.60	126.904	131.293

Platinum	Gold	Mercury	Thallium	Lead	Bismuth	Polonium	Astatine	Radon
78	79	80	81	82	83	84	85	86
Pt	**Au**	**Hg**	**Tl**	**Pb**	**Bi**	**Po**	**At**	**Rn**
195.078	196.967	200.59	204.383	207.2	208.980	(209)	(210)	(222)

Darmstadtium	Roentgenium	Ununbium		Ununquadium
110	111	* 112		* 114
Ds	**Rg**	**Uub**		**Uuq**
(281)	(272)	(285)		(289)

* The names and symbols for elements 112 and 114 are temporary. Final names will be selected when the elements' discoveries are verified.

Europium	Gadolinium	Terbium	Dysprosium	Holmium	Erbium	Thulium	Ytterbium	Lutetium
63	64	65	66	67	68	69	70	71
Eu	**Gd**	**Tb**	**Dy**	**Ho**	**Er**	**Tm**	**Yb**	**Lu**
151.964	157.25	158.925	162.500	164.930	167.259	168.934	173.04	174.967

Americium	Curium	Berkelium	Californium	Einsteinium	Fermium	Mendelevium	Nobelium	Lawrencium
95	96	97	98	99	100	101	102	103
Am	**Cm**	**Bk**	**Cf**	**Es**	**Fm**	**Md**	**No**	**Lr**
(243)	(247)	(247)	(251)	(252)	(257)	(258)	(259)	(262)

Topographic Map Symbols

Topographic Map Symbols

Primary highway, hard surface		Index contour
Secondary highway, hard surface		Supplementary contour
Light-duty road, hard or improved surface		Intermediate contour
Unimproved road		Depression contours
Railroad: single track		
Railroad: multiple track		Boundaries: national
Railroads in juxtaposition		State
		County, parish, municipal
Buildings		Civil township, precinct, town, barrio
Schools, church, and cemetery		Incorporated city, village, town, hamlet
Buildings (barn, warehouse, etc.)		Reservation, national or state
Wells other than water (labeled as to type)		Small park, cemetery, airport, etc.
Tanks: oil, water, etc. (labeled only if water)		Land grant
Located or landmark object; windmill		Township or range line, U.S. land survey
Open pit, mine, or quarry; prospect		Township or range line, approximate location
Marsh (swamp)		
Wooded marsh		Perennial streams
Woods or brushwood		Elevated aqueduct
Vineyard		Water well and spring
Land subject to controlled inundation		Small rapids
Submerged marsh		Large rapids
Mangrove		Intermittent lake
Orchard		Intermittent stream
Scrub		Aqueduct tunnel
Urban area		Glacier
		Small falls
x7369 Spot elevation		Large falls
670 Water elevation		Dry lake bed

Cómo usar el glosario en español:
1. Busca el término en inglés que desees encontrar.
2. El término en español, junto con la definición, se encuentran en la columna de la derecha.

Pronunciation Key

Use the following key to help you sound out words in the glossary.

a	back (BAK)	ew	food (FEWD)
ay	day (DAY)	yoo	pure (PYOOR)
ah	father (FAH thur)	yew	few (FYEW)
ow	flower (FLOW ur)	uh	comma (CAH muh)
ar	car (CAR)	u (+ con)	rub (RUB)
e	less (LES)	sh	shelf (SHELF)
ee	leaf (LEEF)	ch	nature (NAY chur)
ih	trip (TRIHP)	g	gift (GIHFT)
i (i + con + e)	idea (i DEE uh)	j	gem (JEM)
oh	go (GOH)	ing	sing (SING)
aw	soft (SAWFT)	zh	vision (VIH zhun)
or	orbit (OR buht)	k	cake (KAYK)
oy	coin (COYN)	s	seed, cent (SEED, SENT)
oo	foot (FOOT)	z	zone, raise (ZOHN, RAYZ)

English	A	Español

abrasion: a type of erosion that occurs when wind-blown sediments strike rocks and sediments, polishing and pitting their surface. (p. 76)

absolute age: age, in years, of a rock or other object; can be determined by using properties of the atoms that make up materials. (p. 139)

aquifer (AK wuh fur): layer of permeable rock that allows water to flow through. (p. 104)

abrasión: tipo de erosión que ocurre cuando los sedimentos arrastrados por el viento golpean las rocas y los sedimentos, puliendo y llenando de hoyos su superficie. (p. 76)

edad absoluta: edad, en años, de una roca u otro objeto; puede determinarse utilizando las propiedades de los átomos de los materiales. (p. 139)

acuífero: capa de roca permeable que permite que el agua fluya a través de ella. (p. 104)

B

beach: deposit of sediment whose materials vary in size, color, and composition and is most commonly found on a smooth, gently sloped shoreline. (p. 111)

playa: depósito de sedimentos cuyos materiales varían en tamaño, color y composición y que comúnmente se encuentran en las líneas costeras planas y poco inclinadas. (p. 111)

C

carbon film: thin film of carbon residue preserved as a fossil. (p. 126)

cast: a type of body fossil that forms when crystals fill a mold or sediments wash into a mold and harden into rock. (p. 127)

película de carbono: capa delgada de residuos de carbono preservada como un fósil. (p. 126)

vaciado: tipo de cuerpo fósil que se forma cuando los cristales llenan un molde o los sedimentos son lavados hacia un molde y se endurecen convirtiéndose en roca. (p. 127)

cave: underground opening that can form when acidic groundwater dissolves limestone. (p. 107)

Cenozoic (seh nuh ZOH ihk) Era: era of recent life that began about 66 million years ago and continues today; includes the first appearance of *Homo sapiens* about 400,000 years ago. (p. 174)

channel: groove created by water moving down the same path. (p. 94)

chemical weathering: occurs when chemical reactions dissolve the minerals in rocks or change them into different minerals. (p. 39)

climate: average weather pattern in an area over a long period of time; can be classified by temperature, humidity, precipitation, and vegetation. (p. 40)

conic projection: map made by projecting points and lines from a globe onto a cone. (p. 19)

contour farming: planting along the natural contours of the land to reduce soil ersosion. (p. 53)

contour line: line on a map that connects points of equal elevation. (p. 20)

creep: a type of mass movement in which sediments move down-slope very slowly; is common in areas of freezing and thawing, and can cause walls, trees, and fences to lean downhill. (p. 66)

cyanobacteria: chlorophyll-containing, photosynthetic bacteria thought to be one of Earth's earliest life-forms. (p. 163)

cueva: apertura subterránea que puede formarse cuando el agua subterránea acidificada disuelve la piedra caliza. (p. 107)

Era Cenozoica: era de vida reciente que comenzó hace aproximadamente 66 millones de años y continúa hasta hoy; incluye la aparición del *Homo sapiens* cerca de 400,000 años atrás. (p. 174)

canal: surco creado por el agua cuando se mueve cuesta abajo por el mismo curso. (p. 94)

erosión química: ocurre cuando las reacciones químicas disuelven los minerales en las rocas o los convierten en diferentes minerales. (p. 39)

clima: modelo meteorológico promedio en un área durante un periodo de tiempo largo; puede clasificarse por temperatura, humedad, precipitación y vegetación. (p. 40)

proyección cónica: mapa hecho por la proyección de puntos y líneas desde un globo a un cono. (p. 19)

cultivo de contorno: plantación a lo largo de los contornos naturales de la tierra para reducir la erosión de los suelos. (p. 53)

curva de nivel: línea en un mapa que conecta puntos de la misma elevación. (p. 20)

reptación: tipo de movimiento en masa en el que los sedimentos se mueven hacia abajo muy lentamente; es común en áreas sujetas a congelación y descongelación y puede causar que los muros, los árboles y los cercos se inclinen hacia abajo. (p. 66)

cianobacteria: bacteria fotosintética que contiene clorofila; se cree que es una de las primeras formas de vida que surgió en la tierra. (p. 163)

D

deflation: a type of erosion that occurs when wind blows over loose sediments, removes small particles, and leaves coarser sediments behind. (p. 76)

deposition: dropping of sediments that occurs when an agent of erosion, such as gravity, a glacier, wind, or water, loses its energy and can no longer carry its load. (p. 65)

drainage basin: land area from which a river or stream collects runoff. (p. 96)

dune (DOON): mound formed when windblown sediments pile up behind an obstacle; common landform in desert areas. (p. 79)

deflación: tipo de erosión que ocurre cuando el viento sopla sobre los sedimentos sueltos, retira partículas pequeñas y deja los sedimentos grandes. (p. 76)

deposición: caída de sedimentos que ocurre cuando un agente erosivo como la gravedad, un glaciar, el viento o el agua, pierde su energía y ya no puede continuar con su carga. (p. 65)

cuenca de drenaje: terreno del que un río o arroyo recolecta sus aguas. (p. 96)

duna: amontonamiento de tierra formado cuando los sedimentos arrastrados por el aire se apilan detrás de un obstáculo; forma de terreno común en las áreas desérticas. (p. 79)

E

eon: longest subdivision in the geologic time scale that is based on the abundance of certain types of fossils and is subdivided into eras, periods, and epochs. (p. 155)

epoch: next-smaller division of geologic time after the period; is characterized by differences in life-forms that may vary regionally. (p. 155)

equator: imaginary line that wraps around Earth at 0° latitude, halfway between the north and south poles. (p. 14)

era: second-longest division of geologic time; is subdivided into periods and is based on major worldwide changes in types of fossils. (p. 155)

erosion: process in which surface materials are worn away and transported from one place to another by agents such as gravity, water, wind, and glaciers. (p. 64)

eón: la más grande subdivisión en la escala del tiempo geológico; se basa en la abundancia de cierto tipo de fósiles y está dividida en eras, periodos y épocas. (p. 155)

época: la siguiente división más pequeña del tiempo geológico después del periodo; está caracterizada por diferencias en las formas de vida que pueden variar regionalmente. (p. 155)

ecuador: línea imaginaria que rodea a la Tierra en el punto de latitud 0°, a la mitad de la distancia entre el polo norte y el polo sur. (p. 14)

era: la segunda división más grande del tiempo geológico; está subdividida en periodos y se basa en cambios mayores en todo el mundo con respecto a los tipos de fósiles. (p. 155)

erosión: proceso mediante el cual los materiales de la superficie son desgastados y transportados de un lugar a otro por agentes como la gravedad, el agua, el viento o los glaciares. (p. 64)

F

fault-block mountains: mountains formed from huge, tilted blocks of rock that are separated from surrounding rocks by faults. (p. 12)

folded mountains: mountains formed when horizontal rock layers are squeezed from opposite sides, causing them to buckle and fold. (p. 11)

fossils: remains, imprints, or traces of prehistoric organisms that can tell when and where organisms once lived and how they lived. (p. 125)

montañas de fallas: montañas formadas por bloques rocosos grandes e inclinados separados de las rocas circundantes por fracturas. (p. 12)

montañas de plegamiento: montañas formadas cuando las capas rocosas horizontales son comprimidas desde lados opuestos, causando que se colapsen y plieguen. (p. 11)

fósiles: restos, huellas o trazas de organismos prehistóricos que pueden informar cuándo, dónde y cómo vivieron tales organismos. (p. 125)

G

geologic time scale: division of Earth's history into time units based largely on the types of life-forms that lived only during certain periods. (p. 154)

geyser: hot spring that erupts periodically and shoots water and steam into the air—for example, Old Faithful in Yellowstone National Park. (p. 107)

glaciers: large, moving masses of ice and snow that change large areas of Earth's surface through erosion and deposition. (p. 69)

escala del tiempo geológico: división de la historia de la Tierra en unidades de tiempo; se basa en los tipos de formas de vida que vivieron sólo durante ciertos periodos. (p. 154)

géiser: aguas termales que erupcionan periódicamente arrojando agua y vapor al aire –por ejemplo, Old Faithful en el Parque Nacional Yellowstone. (p. 107)

glaciares: grandes masas de hielo y nieve en movimiento que cambian extensas áreas de la superficie terrestre a través de la erosión y la deposición. (p. 69)

groundwater: water that soaks into the ground and collects in pores and empty spaces and is an important source of drinking water. (p. 103)

agua subterránea: agua que se difunde en el suelo y se acumula en poros y espacios vacíos siendo una fuente importante de agua potable. (p. 103)

H

half-life: time it takes for half the atoms of an isotope to decay. (p. 140)

horizon: each layer in a soil profile—horizon A (top layer of soil), horizon B (middle layer), and horizon C (bottom layer). (p. 44)

humus (HYEW mus): dark-colored, decayed organic matter that supplies nutrients to plants and is found mainly in topsoil. (p. 44)

vida media: tiempo que le toma a la mitad de los átomos de un isótopo para desintegrarse. (p. 140)

horizonte: cada capa en un perfil de suelos: horizonte A (la capa superior del suelo), horizonte B (la capa media) y horizonte C (la capa inferior). (p. 44)

humus: materia orgánica en descomposición, de color oscuro, que suministra nutrientes a las plantas y se encuentra principalmente en la parte superior del suelo. (p. 44)

I

ice wedging: mechanical weathering process that occurs when water freezes in the cracks of rocks and expands, causing the rock to break apart. (p. 38)

impermeable: describes materials that water cannot pass through. (p. 104)

index fossils: remains of species that existed on Earth for a relatively short period of time, were abundant and widespread geographically, and can be used by geologists to assign the ages of rock layers. (p. 129)

gelifracción: proceso de erosión mecánica que ocurre cuando el agua se congela en las grietas de las rocas y luego se expande, causando que la roca de fraccione. (p. 38)

impermeable: describe materiales que impiden el paso del agua a través de ellos. (p. 104)

fósiles índice: restos de especies que existieron sobre la Tierra durante un periodo de tiempo relativamente corto y que fueron abundantes y ampliamente diseminadas geográficamente; los geólogos pueden usarlos para inferir las edades de las capas rocosas. (p. 129)

L

latitude: distance in degrees north or south of the equator. (p. 14)

leaching: removal of minerals that have been dissolved in water. (p. 45)

litter: twigs, leaves, and other organic matter that help prevent erosion and hold water and may eventually be changed into humus by decomposing organisms. (p. 45)

loess (LES): windblown deposit of tightly packed, fine-grained sediments. (p. 79)

longitude: distance in degrees east or west of the prime meridian. (p. 15)

latitud: distancia en grados al norte o sur del ecuador. (p. 14)

lixiviación: remoción de minerales que han sido disueltos en el agua. (p. 45)

hojarasca: ramitas, hojas y demás material orgánico que ayuda a prevenir la erosión y a mantener el agua, y que eventualmente puede ser transformado en humus por los organismos descomponedores. (p. 45)

loes: depósito arrastrado por el viento que se compone de sedimentos de partículas finas y se encuentran muy compactados. (p. 79)

longitud: distancia en grados al este u oeste del meridiano inicial. (p. 15)

longshore current: current that runs parallel to the shoreline, is caused by waves colliding with the shore at slight angles, and moves tons of loose sediment. (p. 110)

corriente costera: corriente que corre paralela a la línea costera, es causada por olas que chocan contra la orilla en ángulos tenues y mueve toneladas de sedimentos sueltos. (p. 110)

M

map legend: explains the meaning of symbols used on a map. (p. 22)

map scale: relationship between distances on a map and distances on Earth's surface that can be represented as a ratio or as a small bar divided into sections. (p. 22)

mass movement: any type of erosion that occurs as gravity moves materials down-slope. (p. 65)

meander (mee AN dur): broad, c-shaped curve in a river or stream, formed by erosion of its outer bank. (p. 97)

mechanical weathering: physical processes that break rock apart without changing its chemical makeup; can be caused by ice wedging, animals, and plant roots. (p. 37)

Mesozoic (mez uh ZOH ihk) Era: middle era of Earth's history, during which Pangaea broke apart, dinosaurs appeared, and reptiles and gymnosperms were the dominant land life-forms. (p. 170)

mold: a type of body fossil that forms in rock when an organism with hard parts is buried, decays or dissolves, and leaves a cavity in the rock. (p. 127)

moraine: large ridge of rocks and soil deposited by a glacier when it stops moving forward. (p. 71)

leyenda del mapa: explica el significado de los símbolos utilizados en un mapa. (p. 22)

escala del mapa: relación entre las distancias en un mapa y las distancias sobre la superficie terrestre, que puede representarse como una relación o como una barra pequeña dividida en secciones. (p. 22)

movimiento en masa: cualquier tipo de erosión que ocurre cuando la gravedad mueve materiales cuesta abajo. (p. 65)

meandro: curva amplia en forma de C en un río o arroyo, formada por la erosión de su rivera externa. (p. 97)

erosión mecánica: procesos físicos que fraccionan la roca sin cambiar su composición química; puede ser causada por gelifracción, animales y raíces de las plantas. (p. 37)

Era Mesozoica: era media de la historia de la Tierra durante la cual se escindió la Pangea y aparecieron los dinosaurios; los reptiles y gimnospermas fueron las formas de vida que dominaron la tierra. (p. 170)

moldura: tipo de cuerpo fósil que se formó en la roca cuando un organismo con partes duras fue enterrado, descompuesto o disuelto, dejando una cavidad en la roca. (p. 127)

morrena: grandes cúmulos de rocas y suelo depositados por un glaciar cuando deja de moverse hacia adelante. (p. 71)

N

natural selection: process by which organisms that are suited to a particular environment are better able to survive and reproduce than organisms that are not. (p. 157)

no-till farming: method for reducing soil erosion; plant stalks are left in the field after harvesting and the next year's crop is planted within the stalks without plowing. (p. 52)

selección natural: proceso mediante el cual los organismos que están adaptados a un ambiente particular están mejor capacitados para sobrevivir y reproducirse que los organismos que no están adaptados. (p. 157)

cultivo sin labranza: método para reducir la erosión del suelo; los tallos de las plantas se dejan en el terreno después de la cosecha y el cultivo del siguiente año se siembra entre los tallos sin hacer labranza alguna. (p. 52)

O

organic evolution: change of organisms over geologic time. (p. 156)

outwash: material deposited by meltwater from a glacier. (p. 71)

oxidation (ahk sih DAY shun): chemical weathering process that occurs when some minerals are exposed to oxygen and water over time. (p. 40)

evolución orgánica: cambio de los organismos a través del tiempo geológico. (p. 156)

derrubio: material depositado por la corriente de agua del hielo derretido de un glaciar. (p. 71)

oxidación: proceso de erosión química que ocurre cuando algunos minerales son expuestos al oxígeno y al agua. (p. 40)

P

Paleozoic Era: era of ancient life, which began about 544 million years ago, when organisms developed hard parts, and ended with mass extinctions about 245 million years ago. (p. 164)

Pangaea (pan JEE uh): large, ancient landmass that was composed of all the continents joined together. (p. 161)

period: third-longest division of geologic time; is subdivided into epochs and is characterized by the types of life that existed worldwide. (p. 155)

permeable (PUR mee uh bul): describes soil and rock with connecting pores through which water can flow. (p. 104)

permineralized remains: fossils in which the spaces inside are filled with minerals from groundwater. (p. 126)

plain: large, flat landform that often has thick, fertile soil and is usually found in the interior region of a continent. (p. 8)

plateau (pla TOH): flat, raised landform made up of nearly horizontal rocks that have been uplifted. (p. 10)

plucking: process that adds gravel, sand, and boulders to a glacier's bottom and sides as water freezes and thaws, breaking off pieces of surrounding rock. (p. 70)

Precambrian (pree KAM bree un) time: longest part of Earth's history, lasting from 4.0 billion to about 544 million years ago. (p. 162)

prime meridian: imaginary line that represents 0° longitude and runs from the north pole through Greenwich, England, to the south pole. (p. 15)

Era Paleozoica: era de la vida antigua que comenzó hace 544 millones de años, cuando los organismos desarrollaron partes duras; terminó con extinciones en masa hace unos 245 millones de años. (p. 164)

Pangea: masa terrestre antigua que estaba compuesta por todos los continentes unidos. (p. 161)

periodo: la tercera división más grande del tiempo geológico; está subdividido en épocas y se caracteriza por los tipos de vida que existieron en todo el mundo. (p. 155)

permeable: describe el suelo y la roca con poros conectados a través de los cuales el agua puede fluir. (p. 104)

restos permineralizados: fósiles en los que los espacios interiores son llenados con minerales de aguas subterráneas. (p. 126)

planicie: formación de terreno extenso y plano que a menudo tiene suelos gruesos y fértiles; generalmente se encuentra en la región interior de un continente. (p. 8)

meseta: formación de terreno plano y elevado constituida por rocas casi horizontales que han sido levantadas. (p. 10)

gelivación: proceso que agrega grava, arena y cantos a la parte inferior y lateral de un glaciar conforme el agua se congela y descongela, fraccionando las piezas de las rocas circundantes. (p. 70)

tiempo precámbrico: la parte más duradera de la historia de la Tierra; duró desde hace 4.0 billones de años hasta hace aproximadamente 544 millones de años. (p. 162)

meridiano inicial: línea imaginaria que representa los cero grados de longitud y va desde el polo norte pasando por Greenwich, Inglaterra, hasta el polo sur. (p. 15)

principle of superposition: states that in undisturbed rock layers, the oldest rocks are on the bottom and the rocks become progressively younger toward the top. (p. 132)

principio de superposición: establece que en las capas rocosas no perturbadas, las rocas más antiguas están en la parte inferior y las rocas son más jóvenes conforme están más cerca de la superficie. (p. 132)

R

radioactive decay: process in which some isotopes break down into other isotopes and particles. (p. 139)

radiometric dating: process used to calculate the absolute age of rock by measuring the ratio of parent isotope to daughter product in a mineral and knowing the half-life of the parent. (p. 141)

relative age: the age of something compared with other things. (p. 133)

runoff: any rainwater that does not soak into the ground or evaporate but flows over Earth's surface; generally flows into streams and has the ability to erode and carry sediments. (p. 92)

desintegración radiactiva: proceso en el que algunos isótopos se desintegran en otros isótopos y partículas. (p. 139)

fechado radiométrico: proceso utilizado para calcular la edad absoluta de las rocas midiendo la relación isótopo parental a producto derivado en un mineral y conociendo la vida media del parental. (p. 141)

edad relativa: la edad de algo comparado con otras cosas. (p. 133)

escorrentía: agua de lluvia que no se difunde en el suelo ni se evapora pero que fluye sobre la superficie terrestre; generalmente fluye hacia los arroyos y tiene la capacidad de causar erosión y transportar sedimentos. (p. 92)

S

sheet erosion: a type of surface water erosion caused by runoff that occurs when water flowing as sheets picks up sediments and carries them away. (p. 95)

slump: a type of mass movement that occurs when a mass of material moves down a curved slope. (p. 65)

soil: mixture of weathered rock and mineral fragments, decayed organic matter, water, and air that can take thousands of years to develop. (p. 42)

soil profile: vertical section of soil layers, each of which is a horizon. (p. 44)

species: group of organisms that reproduces only with other members of their own group. (p. 156)

spring: forms when the water table meets Earth's surface; often found on hillsides and used as a freshwater source. (p. 107)

erosión laminar: tipo de erosión causada por las corrientes de agua de lluvia; ocurre cuando el agua fluye laminarmente recogiendo sedimentos y llevándolos a otro lugar. (p. 95)

desprendimiento: tipo de movimiento en masa que ocurre cuando un volumen de material se mueve hacia abajo de una cuesta curvada. (p. 65)

suelo: mezcla de roca erosionada y fragmentos minerales, materia orgánica en descomposición, agua y aire, y que puede tardar miles de años para formarse. (p. 42)

perfil de suelos: sección vertical de las capas del suelo, cada una de las cuales es un horizonte. (p. 44)

especie: grupo de organismos que se reproduce sólo entre los miembros de su mismo grupo. (p. 156)

manantial: se forma cuando el nivel freático alcanza la superficie terrestre; a menudo se encuentran en las laderas y se usan como fuente de agua potable. (p. 107)

T

terracing: farming method used to reduce erosion on steep slopes. (p. 53)

terraceo: método de siembra usado para reducir la erosión en cuestas inclinadas. (p. 53)

till: mixture of different-sized sediments that is dropped from the base of a retreating glacier and can cover huge areas of land. (p. 70)

topographic map: map that shows the changes in elevation of Earth's surface and indicates such features as roads and cities. (p. 20)

trilobite (TRI luh bite): organism with a three-lobed exoskeleton that was abundant in Paleozoic oceans and is considered to be an index fossil. (p. 155)

tillita: mezcla de sedimentos de diferentes tamaños que ha caído de la base de un glaciar en retroceso y puede cubrir grandes extensiones de terreno. (p. 70)

mapa topográfico: mapa que muestra los cambios en la elevación de la superficie terrestre que puede ser representado como una relación e indica características como carreteras y ciudades. (p. 20)

trilobite: organismo con un exoesqueleto trilobulado que fue abundante en los océanos del Paleozoico y es considerado como un fósil índice. (p. 155)

unconformity (un kun FOR mih tee): gap in the rock layer that is due to erosion or periods without any deposition. (p. 134)

uniformitarianism: principle stating that Earth processes occurring today are similar to those that occurred in the past. (p. 143)

upwarped mountains: mountains formed when blocks of Earth's crust are pushed up by forces inside Earth. (p. 12)

discordancia: brecha en la capa rocosa que es debida a la erosión o a periodos sin deposición. (p. 134)

uniformitarianismo: principio que establece que los procesos de la Tierra que ocurren actualmente son similares a los que ocurrieron en el pasado. (p. 143)

montañas de levantamiento: montañas que se forman cuando los bloques de la corteza terrestre son empujados hacia arriba por fuerzas del interior de la Tierra. (p. 12)

volcanic mountains: mountains formed when molten material reaches Earth's surface through a weak crustal area and piles up into a cone-shaped structure. (p. 13)

montañas volcánicas: montañas formadas cuando material derretido alcanza la superficie a través de un área débil de la corteza terrestre y se acumula formando una estructura en forma de cono. (p. 13)

water table: upper surface of the zone of saturation; drops during a drought. (p. 104)

weathering: mechanical or chemical surface processes that break rock into smaller and smaller pieces. (p. 36)

nivel freático: parte superior de la zona de saturación; desciende durante las sequías. (p. 104)

erosión: proceso superficial químico o mecánico que fracciona la roca en trozos cada vez más pequeños. (p. 36)

Italic numbers = illustration/photo **Bold numbers** = vocabulary term
lab = a page on which the entry is used in a lab
act = a page on which the entry is used in an activity

Index

Index

Index

Magnification Key: Magnifications listed are the magnifications at which images were originally photographed.
LM–Light Microscope
SEM–Scanning Electron Microscope
TEM–Transmission Electron Microscope

Acknowledgments: Glencoe would like to acknowledge the artists and agencies who participated in illustrating this program: Absolute Science Illustration; Andrew Evansen; Argosy; Articulate Graphics; Craig Attebery, represented by Frank & Jeff Lavaty; CHK America; John Edwards and Associates; Gagliano Graphics; Pedro Julio Gonzalez, represented by Melissa Turk & The Artist Network; Robert Hynes, represented by Mendola Ltd.; Morgan Cain & Associates; JTH Illustration; Laurie O'Keefe; Matthew Pippin, represented by Beranbaum Artist's Representative; Precision Graphics; Publisher's Art; Rolin Graphics, Inc.; Wendy Smith, represented by Melissa Turk & The Artist Network; Kevin Torline, represented by Berendsen and Associates, Inc.; WILDlife ART; Phil Wilson, represented by Cliff Knecht Artist Representative; Zoo Botanica.

Photo Credits

Cover Jack Dykinga/Getty Images; **i ii** Jack Dykinga/Getty Images; **iv** (bkgd)John Evans, (inset)Jack Dykinga/Getty Images; **v** (t)PhotoDisc, (b)John Evans; **vi** (l)John Evans, (r)Geoff Butler; **vii** (l)John Evans, (r)PhotoDisc; **viii** PhotoDisc; **ix** Aaron Haupt Photography; **x** Sylvester Allred/Visuals Unlimited; **xi** (t)PhotoTake NYC/PictureQuest, (b)Curt Schieber; **xii** John Giustina/Getty Images; **2** Charles Palek/Earth Scenes; **3** (t)Tom Bean/Stone, (b)JC Marchak/AP/Wide World Photos; **4** C. C. Lockwood/Earth Scenes; **5** (t)David Ulmer/Stock Boston, (b)Mark Burnett; **6–7** GSFC/NASA; **9** (tl)Alan Maichrowicz/Peter Arnold, Inc., (tr)Carr Clifton/Minden Pictures, (b)Stephen G. Maka/DRK Photo; **10** Ron Mellot; **11** John Lemker/Earth Scenes; **12** (t)John Kieffer/Peter Arnold, Inc., (b)Carr Clifton/Minden Pictures; **13** David Muench/CORBIS; **16** Dominic Oldershaw; **21** (t)Rob & Ann Simpson, (b)courtesy Maps a la Carte, Inc. and TopoZone.com; **24** CORBIS; **26** (t)Layne Kennedy/CORBIS, (b)John Evans; **27** John Evans; **28** (tl)Culver Pictures, (tcl)PhotoDisc, (tcr)William Manning/The Stock Market/CORBIS, (tr)Kunio Owaki/The Stock Market/CORBIS, (c)Pictor, (b)PhotoDisc; **29** (l)Tom Bean/DRK Photo, (r)Marc Muench; **30** William Weber; **32** Aaron Haupt; **34–35** Andrew Brown, Ecoscene/CORBIS; **37** (l)StudiOhio, (r)Tom Bean/DRK Photo; **38** W. Perry Conway/CORBIS; **39** Hans Strand/Stone; **40** (tl)Craig Kramer, (tr)A.J. Copley/Visuals Unlimited, (bl br)John Evans; **41** (l)William Johnson/Stock Boston, (r)Runk/Schoenberger from Grant Heilman; **43** (bkgd)Stephen R. Wagner, (t)James D. Balog, (c)Martin Miller, (b)Steven C. Wilson/Entheos; **44** (l)Bonnie Heidel/Visuals Unlimited, (r)John Bova/Photo Researchers; **50** (l)Gary Braasch/CORBIS, (r)Donna Ikenberry/Earth Scenes; **51** Chip & Jill Isenhart/Tom Stack & Associates; **52** (t)Dr. Russ Utgard, (b)Denny Eilers from Grant Heilman; **53** Georg Custer/Photo Researchers; **54** (t)George H. Harrison from Grant Heilman, (b)Bob

Daemmrich; **55** KS Studios; **56** Larry Hamill; **57** (l)Tom Bean/DRK Photo, (r)David M. Dennis/Earth Scenes; **59** Matt Meadows; **60** Georg Custer/Photo Researchers; **62–63** Paul A. Souders/CORBIS; **64** Robert L. Schuster/USGS; **65** Martin G. Miller/Visuals Unlimited; **66** (t)John D. Cunningham/Visuals Unlimited, (bl)Sylvester Allred/Visuals Unlimited, (br)Tom Uhlman/Visuals Unlimited; **67** AP/Wide World Photos; **68** Martin G. Miller/Visuals Unlimited; **70** James N. Westwater; **71** (t)Tom Bean/Stone/Getty Images, (b)Tom Bean/CORBIS; **72** John Gerlach/Visuals Unlimited; **73** Gregory G. Dimijian/Photo Researchers; **74** Mark E. Gibson/Visuals Unlimited; **75** Timothy Fuller; **76** Galen Rowell/CORBIS; **78** Fletcher & Baylis/Photo Researchers; **79** (t)John D. Cunningham/Visuals Unlimited, (b)file photo; **80** (bkgd)Breck P. Kent/Earth Scenes, (tl)Stephen J. Krasemann/Photo Researchers, (tr)Steve McCurry, (b)Wyman P. Meinzer; **81** John Giustina/FPG/Getty Images; **82** (t)Greg Vaughn/Tom Stack & Assoc., (b)Matt Meadows; **84** (t)World Class Images, (c)Yann Arthus-Bertrand/CORBIS, (b)AP/Wide World Photos; **86** John D. Cunningham/Visuals Unlimited; **90–91** William Manning/The Stock Market/CORBIS; **91** Aaron Haupt; **92** (l)Michael Busselle/Stone/Getty Images, (r)David Woodfall/DRK Photo; **93** Tim Davis/Stone/Getty Images; **94** (t)Grant Heilman Photography, (b)KS Studios; **96** Mel Allen/ICL/Panoramic Images; **98** CORBIS/PictureQuest; **99** (l)Harald Sund/The Image Bank/Getty Images, (r)Loren McIntyre; **100** C. Davidson/Comstock; **101** James L. Amos/CORBIS; **102** (l)Wolfgang Kaehler, (r)Nigel Press/Stone/Getty Images; **103** First Image; **105** CORBIS; **106** file photo; **107** Barbara Filet; **108** Chad Ehlers/Stone/Getty Images; **110** Macduff Everton/The Image Bank/Getty Images; **111** (l)Steve Bentsen, (tr)SuperStock, (bl)Runk/Schoenberger from Grant Heilman, (br)Breck P. Kent/Earth Scenes; **112** Bruce Roberts/Photo Researchers; **114 115** KS Studios; **116** Gary Bogdon/CORBIS Sygma; **117** (l)Todd Powell/Index Stock, (r)J. Wengle/DRK Photo; **120** Barbara Filet; **121** Grant Heilman Photography; **122–123** Hugh Sitton/Getty Images; **124** (t)Mark E. Gibson/Visuals Unlimited, (b)D.E. Hurlbert & James DiLoreto/Smithsonian Institution; **125** Jeffrey Rotman/CORBIS; **126** (t)Dr. John A. Long, (b)A.J. Copley/Visuals Unlimited; **128** (t)PhotoTake, NYC/PictureQuest, (bl br)Louis Psihoyos/Matrix; **130** David M. Dennis; **131** (l)Gary Retherford/Photo Researchers, (r)Lawson Wood/CORBIS; **132** Aaron Haupt; **135** (bkgd)Lyle Rosbotham, (l)IPR/12-18 T. Bain, British Geological Survey/NERC. All rights reserved, (r)Tom Bean/CORBIS; **136** Jim Hughes/PhotoVenture/Visuals Unlimited; **137** (l)Michael T. Sedam/CORBIS, (r)Pat O'Hara/CORBIS; **139** Aaron Haupt; **141** James King-Holmes/Science Photo Library/Photo Researchers; **142** Kenneth Garrett; **143** WildCountry/CORBIS; **144** (t)A.J. Copley/Visuals Unlimited, (b)Lawson Wood/CORBIS; **145** Matt Meadows; **146** Jacques Bredy; **147** (tl)Francois Gohier/Photo Researchers, (tr)Sinclair Stammers/Photo Researchers, (b)Mark E. Gibson/DRK Photo; **150** Tom Bean/CORBIS; **152–153** Roger Garwood & Trish Ainslie/CORBIS; **153** KS Studios; **154** Tom & Therisa Stack/Tom Stack & Assoc.; **156** (l)Gerald & Buff Corsi/Visuals Unlimited, (r)John Gerlach/Animals Animals; **158** (tl)Mark Boulron/Photo Researchers, (others)Walter

Chandoha; **159** Jeff Lepore/Photo Researchers; **163** (l)Mitsuaki Iwago/Minden Pictures, (r)R. Calentine/ Visuals Unlimited; **165** J.G. Gehling/Discover Magazine; **166** Gerry Ellis/ENP Images; **169** Matt Meadows; **172** (l)David Burnham/Fossilworks, Inc., (r)François Gohier/Photo Researchers; **174** Michael Andrews/Earth Scenes; **175** Tom J. Ulrich/Visuals Unlimited; **176** David M. Dennis; **177** Mark Burnett; **179** (l)E. Webber/Visuals Unlimited, (r)Len Rue, Jr./Animals Animals; **181** John Cancalosi/Stock Boston; **184** PhotoDisc; **186** Tom Pantages; **190** Michell D. Bridwell/PhotoEdit, Inc.; **191** (t)Mark Burnett, (b)Dominic Oldershaw; **192** StudioOhio; **193** Timothy Fuller; **194** Aaron Haupt; **196** KS Studios; **197** Matt Meadows; **198** PhotoDisc; **199** Giboux/Getty Images; **201** Amanita Pictures; **202** Bob Daemmrich; **204** Davis Barber/PhotoEdit, Inc.

PERIODIC TABLE OF THE ELEMENTS

Columns of elements are called groups. Elements in the same group have similar chemical properties.

Gas
Liquid
Solid
Synthetic

Element — Hydrogen
Atomic number — 1
Symbol — H
Atomic mass — 1.008
State of matter

The first three symbols tell you the state of matter of the element at room temperature. The fourth symbol identifies elements that are not present in significant amounts on Earth. Useful amounts are made synthetically.

Group	1	2	3	4	5	6	7	8	9
1	Hydrogen 1 H 1.008								
2	Lithium 3 Li 6.941	Beryllium 4 Be 9.012							
3	Sodium 11 Na 22.990	Magnesium 12 Mg 24.305							
4	Potassium 19 K 39.098	Calcium 20 Ca 40.078	Scandium 21 Sc 44.956	Titanium 22 Ti 47.867	Vanadium 23 V 50.942	Chromium 24 Cr 51.996	Manganese 25 Mn 54.938	Iron 26 Fe 55.845	Cobalt 27 Co 58.933
5	Rubidium 37 Rb 85.468	Strontium 38 Sr 87.62	Yttrium 39 Y 88.906	Zirconium 40 Zr 91.224	Niobium 41 Nb 92.906	Molybdenum 42 Mo 95.94	Technetium 43 Tc (98)	Ruthenium 44 Ru 101.07	Rhodium 45 Rh 102.906
6	Cesium 55 Cs 132.905	Barium 56 Ba 137.327	Lanthanum 57 La 138.906	Hafnium 72 Hf 178.49	Tantalum 73 Ta 180.948	Tungsten 74 W 183.84	Rhenium 75 Re 186.207	Osmium 76 Os 190.23	Iridium 77 Ir 192.217
7	Francium 87 Fr (223)	Radium 88 Ra (226)	Actinium 89 Ac (227)	Rutherfordium 104 Rf (261)	Dubnium 105 Db (262)	Seaborgium 106 Sg (266)	Bohrium 107 Bh (264)	Hassium 108 Hs (277)	Meitnerium 109 Mt (268)

The number in parentheses is the mass number of the longest-lived isotope for that element.

Rows of elements are called periods. Atomic number increases across a period.

The arrow shows where these elements would fit into the periodic table. They are moved to the bottom of the table to save space.

Lanthanide series	Cerium 58 Ce 140.116	Praseodymium 59 Pr 140.908	Neodymium 60 Nd 144.24	Promethium 61 Pm (145)	Samarium 62 Sm 150.36
Actinide series	Thorium 90 Th 232.038	Protactinium 91 Pa 231.036	Uranium 92 U 238.029	Neptunium 93 Np (237)	Plutonium 94 Pu (244)